Praise for Operational Intelligence for Health, Wellness, and Leadership

The road map created through *Operational Intelligence for Health, Wellness, and Leadership* is long overdue in Public Safety. I believe the reader of this material will gain a solid understanding of how to live a healthier and happier life so they may serve the citizens they are sworn to protect.

> —Rick Best, NFFF National Advocate Manager; and retired Firefighter for Westerville Fire Department, Ohio.

As Drs. Baer and Schary promise, their book fills a gap vital to the well-being of firefighters by addressing the operational level of health and wellness, which exists between the strategic and tactical levels. To address this level, the authors offer a curriculum to help firefighters understand "what is involved in health, wellness, and leadership," and what is meant by health, wellness, and leadership. That they do this as a gift to the profession, without seeking personal profit, makes this book all the more worthy of promotion and praise

> —Rev. Christopher Dreisbach, PhD, Associate Director, Organizational Leadership, Johns Hopkins University; and Interim Priest, All Saints Episcopal Church, New York City.

Operational Intelligence for Health, Wellness, and Leadership is an impressive, straightforward, guide for individual and organizational growth. This book will help individuals, groups, and entire fire departments look within for legacy growth. For someone who enters the fire service, this book provides a map to choose a life full of understanding for whatever role he or she participates in. It also offers a host of thoughts and knowledge to have a full personal journey apart from public service. The book offers a portal to senior and future leaders to make positive structural impacts in their fire department. The positive impact can be generational to the future leaders of the United States of America and abroad. I know that last sentence is pretty bold, but so are the authors' contributions. A leadership guide that follows the fire service's deep rooted traditions of helping others is in your hands now, literally.

> —Joe Minogue, NFFF Liaison to FDNY; and retired FDNY Lieutenant.

Whether it's in the firehouse, out in the community, or in your home, the foundation this book lays out serves us in all facets of our lives. Further, our profession goes far beyond strategy and tactics and this guide addresses the human elements at play. There is a severe mental, emotional, and physical demand placed upon the fire service profession. These are all domains that must be curated and developed throughout our lifetime. The community we serve and our families at home are deserving of it, and we must depend on it. This book can serve as the foundation to a holistic approach to fire service wellness and resiliency that affects everything we do downstream.

>   —*Joshua Burchick, Captain for Howard County Fire and Rescue, Maryland.*

The mission of the National Fallen Firefighters Foundation is to honor America's fallen fire heroes, support their families, colleagues, and organizations, and work to reduce preventable firefighter death and injury. Reducing line-of-duty deaths is paramount to the mission and something the organization has strived to do for over two decades. The development of mental health programs, fitness and cardio vascular disease prevention, and cancer prevention has been at the forefront of the NFFF and the First Responder Center of Excellence's efforts. This book educates the reader on all aspects of firefighter health and wellness. Just like turnout gear, this book is an ensemble that all fits together to protect the firefighter. So it is with each individual firefighter; the physical, mental, and spiritual well-being comprises the ensemble that allows us to perform optimally to serve our communities. I'm grateful for the work that's gone into developing this book. It speaks to the integration of the whole human and all the aspects that make us complete, rather than a siloed approach.

>   —*Victor Stagnaro, NFFF Executive Director; and retired Deputy Chief for Prince Georges County Fire Department, Maryland.*

As fire chief, I am passionate about the importance of wellness, health, and total body and mind fitness for firefighters and all first responders. We face unique challenges in our profession, and it is essential that we take care of ourselves physically and mentally in order to be at our best when responding to emergencies. This book is an important resource for anyone who wants to learn more about how to achieve total wellness. It covers a wide range of topics, including physical fitness, nutrition, sleep hygiene, stress management, and mental health. Dr. Baer provides practical advice and strategies that readers can implement in their own lives. I highly recommend this book to firefighters, first responders, and anyone else who wants to live a healthier and more fulfilling life. By taking care of ourselves physically and mentally, we can better serve our communities and be there for our loved ones.

>   —*John Butler, Fire Chief for Fairfax County Fire and Rescue, Virginia; and retired Gunnery Sergeant, USMC.*

This book is an essential resource for current and emerging leaders in the fire service. It presents a comprehensive and research-based approach to addressing the multifaceted challenges of behavioral health, wellness, and leadership in the fire service. From the highest levels of decision-making in the boardroom to the everyday dynamics of the fire station, kitchen table conversations, and critical tailboard discussions, this book equips firefighters with a framework to cultivate a culture of wellbeing and resilience. Thank you Dr. Baer and Dr. Schary for your dedication and hard work.

>   —Robert Perez, Firefighter/Paramedic for Los Angeles Fire Department, California; and NFFF Advocate for LAFD.

For people looking to make a difference in their organization, consider this book. Countless change management theories are offered, but few hit the mark. Those that come close provide little guidance and are only conditionally effective. In *Operational Intelligence for Health, Wellness, and Leadership*, Baer and Schary peel back critical layers for current, and would-be, leaders, as they lay out what is needed to create and sustain positive change in themselves as well as the people they are responsible for leading!

>   —Moses Jefferies IV, Chief of Training for Nashville Fire Department, Tennessee; and NFFF Advocate and Fellow.

How do firefighters die? This landmark book answers that question and provides practical models for combating three of our greatest threats: heart attacks, cancer, and suicide. Through in-depth research and a philosophical framework, *Operational Intelligence for Health, Wellness, and Leadership* seeks to bridge the gap between knowledge and implementation in day-to-day operations. It is a stark treatise on the pitfalls in our field and how we knowingly or unknowingly contribute to our demise. This book should be mandatory reading for everyone in the fire service, from recruits to individuals at the highest levels of leadership. Knowledge is how we can save those who save others, starting with ourselves.

>   —Dustin Zamboni, Fire Marshal for Salt River Pima-Maricopa, Arizona; Instructor at NFA; and NFFF Advocate.

A fire chief's duty is to safeguard lives, property and the community through strategic leadership and effective operations. One powerful tool we can provide to our troops is education. Discover a transformative journey within the pages of *Operational Intelligence for Health, Wellness, and Leadership*. With deep insights into health, wellness, and leadership, Dr. Gamaliel Baer and Dr. David Schary empower readers to unlock their full potential. This compelling guide is a testament to the commitment and passion for fostering balance within our departments to help us thrive. Dive in and embark on a path to professional and personal excellence!
—Joanne Rund, Fire Chief for Baltimore County Fire Department, Maryland; and Region III Coordinator for NFFF Advocates.

Operational Intelligence for Health, Wellness, and Leadership

OIHWL

OPERATIONAL INTELLIGENCE FOR HEALTH, WELLNESS, AND LEADERSHIP

GAMALIEL BAER // DAVID SCHARY

FOREWORD BY FRANK RICCI

Fire Engineering
BOOKS & VIDEOS

Disclaimer
The recommendations, advice, descriptions, and methods in this book are presented solely for educational purposes. The author and publisher assume no liability whatsoever for any loss or damage that results from the use of any of the material in this book. Use of the material in this book is solely at the risk of the user.

Copyright © 2024 by
Fire Engineering Books & Videos
110 S. Hartford Ave., Suite 200
Tulsa, Oklahoma 74120 USA

800.752.9764
+1.918.831.9421
info@fireengineeringbooks.com
www.FireEngineeringBooks.com

Senior Vice President: Eric Schlett
Vice President: Amanda Champion
Operations Manager: Holly Fournier
Sales Manager: Joshua Neal
Managing Editor: Mark Haugh
Production Manager: Tony Quinn
Developmental Editor: Chris Barton
Cover Designer: Brandon Ash
Book Designer: Robert Kern, TIPS Publishing Services, Carrboro, NC

Library of Congress Cataloging-in-Publication Data Available on Request

ISBN: 9781593705886

All rights reserved. No part of this book may be reproduced, stored in a retrieval system, or transcribed in any form or by any means, electronic or mechanical, including photocopying and recording, without the prior written permission of the publisher.

Published in the United States of America

1 2 3 4 5   28 27 26 25 24

CONTENTS

Foreword... vii
Preface... xi
Acknowledgments... xiii
Introduction.. xvii

1   Resilience... 1
    Why is Health and Wellness Important?...................... 2
    Defining Health and Wellness............................... 2
    Defining Resilience.. 3
    Growing Human Resilience................................... 6
    A Need for Health and Wellness Education................... 7
    Individual and Organizational Responsibility.............. 11
    The Change Assessment Model............................... 14
    Micro: You and The Cell................................... 18
    Macro: The Group and You.................................. 21
    Conclusion.. 25
    Questions... 25
    Notes... 26

2   Physical Wellness... 31
    Part 1: Nature and Nurture................................ 31
    Part 2: Deficiency and Excess............................. 42
    Part 3: Micro and Macro................................... 63
    Conclusion.. 71
    Questions... 72
    Notes... 73

3 Nutritional Wellness 85
- Introduction 85
- Part 1: Nature and Nurture 85
- Part 2: Deficiency and Excess 90
- Part 3: Micro and Macro 113
- Conclusion 117
- Questions 118
- Notes 118

4 Mental Wellness 131
- Introduction 131
- Part 1: Nature and Nurture 138
- Part 2: Deficiency and Excess 146
- Part 3: Micro and Macro 167
- Conclusion 171
- Questions 172
- Notes 173

5 Leadership 179
- Introduction 179
- Part 1: Nature and Nurture 180
- Part 2: Deficiency and Excess 196
- Part 3: Micro and Macro 214
- Conclusion 225
- Questions 226
- Notes 226

Answers to End of Chapter Questions 229

Index 243

About the Authors 257

FOREWORD

An ailing professor and author of *The Last Lecture*, Randy Pausch, taught us, "When there's an elephant in the room, introduce him." I was not the first choice to write this foreword, but a stand-in—second choice for the late Bobby Halton, who recently passed before he put pen to paper. Bobby saw the value in this book and was excited to have Fire Engineering introduce this material to the fire service.

Bobby was not only passionate about the book; he was ardent about the lead author. It is with a heavy heart that I agreed to complete this endeavor, and I will attempt the daunting task of combining our perspectives based on our conversations. Bobby introduced Dr. Gamaliel "G" Baer to me with a simple text: "G is a brilliant young man, Coast Guard intelligence and firefighter, you will enjoy him."

I stand with Bobby. I am confident you will find Dr. Baer and Dr. Schary's book valuable and that it will stand with a growing body of work to close the knowledge gap on health and wellness. Perhaps most importantly, this book challenges decades of leadership doctrine and explains how leadership is inherently tied to resilience, health, and wellness.

Using science that can be questioned and tested, this book provides a bridge over a crevasse in the fire service literature that has been lacking an adoptable model curriculum regarding foundational information on health and wellness education.

Any firefighter can read it, and any fire department can teach it. This book is what is necessary—even if another book eventually displaces this one. If a fire department wants to deliver their own health and wellness education—especially on heart attack, cancer, and suicide risk reduction—then right now, this is the only fire service book that provides a place to start.

The authors look to ancient wisdom, as it is the gold standard because it has withstood the test of time and thus is proven, durable, and worthy. To do otherwise is immature and pointless. Dr. Baer and Dr. Schary share Aristotle's philosophy in a way to address behavior change. This is one of the reasons that Bobby and I found value in this book.

Aristotle once said, "Courage is the mother of all virtues because without it, you cannot consistently perform the others." Aristotle is not just talking about physical courage. And, as you would expect, there is no shortage of firefighters who would risk their life to save another. Most firefighters will risk their life at a fire without a second thought. Unfortunately, they will also risk their health and wellness by giving into external influences, bad food, poor nutrition, and not enough or inconsistent exercise, turning away from the light of knowledge because it is hard and not as much fun.

Many of those same firefighters wouldn't risk their career or potential ridicule, or have the courage to close the knowledge gap or self-reflect without lying to themselves. Dr. Baer and Dr. Schary demonstrate that courage in writing this book, with a call for the fire service to look closer at its lifestyle behaviors for answers to health and wellness issues.

For true growth to occur, we all need to take a sober look with an open mind, even if you have a different perspective. This book explains the connection between a person's wellness efforts and their risk factors for heart attack, cancer, and suicide. These issues have been brought to the forefront of the fire service for occupational factors, but addressing the lifestyle factors has fallen short.

This threw me off at first when I read the draft manuscript that you now hold in your hand. I questioned G, "Where is the text and focus on the external factors that we know affect our health?" I went down the litany of environmental factors ranging from the chemical cocktail of smoke that we are all exposed to, heat stress from our protective equipment, known carcinogens, and untested, unregulated forever chemicals in our gear, along with the all-too-common crutch of stress and PTSD without a focus of post-traumatic growth.

Dr. Baer rebutted each one of my concerns with valid points that I was forced to consider, and he did so without dismissing environmental impacts, which he acknowledges exist. We all seem to be moving in the right direction of addressing environmental concerns while almost ignoring the larger issues that are controllable and that only compound the inherent risks of the job. G was introducing me to the "elephant in the room."

All of a sudden, I understood the modern-day philosopher that Bobby saw in G. Philosophy is one of the things that connected Bobby, G, and me. Gandhi said, "If you want to change the world, start with yourself." But for my career, I used bar science and firefighter logic. I could go through two SCBA bottles and outwork the younger guys—hell, I could come to work hungover and bury the recruits on a three-mile run. I called it functional fitness; a Clydesdale horse who could pull his weight. I lived this way all while not having the courage to be honest with myself that I was not healthy. After all, "Fat, drunk, and stupid is no way to go through life, son." Since reading this book, I have started a new page of wellness efforts in my own life.

I've had too many moments in the fire house when it's not fun anymore. When the silence of inaction after the avoidable death from one or more of society's ills is the loudest thing you can hear. I want to see this type of silence go away, and I believe we can work toward it.

You see, the fire service is a reflection of society and all its ills. When we judge the environmental impacts of our health from the job, we start based off of a false premise. We judge our cancer, heart attack, stroke, and suicide rates compared to the general public.

Does anyone believe the general public is healthy overall, or has society traded good health for good medication?

The authors use philosophy and science as knowledge to bring action and change in its purest form, free from influence. Dr. Baer and Dr. Schary give us no place to hide from our reality. While every state and a growing number of departments have made progress addressing health and wellness issues, the changes have been limited in making a substantial impact to reduce lifestyle health and wellness issues, especially heart attacks, cancer, and suicide.

I'm honored to present this book by my friend G and Dr. David Schary. They are moving the fire service forward to a body, mind, and soul approach. They give us the knowledge to do this free of coercion and in a nonpunitive way that doesn't leave us in the dark, but that provides us a light. Aristotle taught us, "Knowing yourself is the beginning of all wisdom." It is from knowing ourselves that we know we can all do better.

> In Liberty,
> Frank Ricci
> Fellow at a Think Tank
> Ret. Battalion Chief New Haven, CT
> Lead Plaintiff in Landmark SCOTUS Case
> FDIC & Fire Engineering Advisory Board
> Author of *Command Presence*

PREFACE

This book has two main goals. The first and overarching goal is to provide the fire service a model for educating on health and wellness based on the relevant areas of the National Fallen Firefighter Foundation's (NFFF) 16 Firefighter Life Safety Initiatives, the National Fire Protection Association's (NFPA) standards, and the International Association of Firefighters and International Association of Fire Chiefs Wellness-Fitness Initiative (IAFF/IAFC-WFI). This book not only educates on the different areas to navigate about health, wellness, and leadership, but also why correctly navigating those issues is important when it comes to the fire service's fight against heart attack, cancer, and suicide. At the time of publication, such a model did not exist.

The second goal is to provide clarity, using both an evidence-based approach and philosophical reasoning, to the topics of resilience, physical wellness, nutritional wellness, mental wellness, and leadership, as well as how those topics tie together. This first edition of *Operational Intelligence for Health, Wellness, and Leadership* does not include a chapter on spiritual wellness, because while the idea of chaplaincy is covered by both the NFFF (in their Fire Service Behavioral Health Management Guide) and the IAFF/IAFC-WFI, it is frankly still considered a subtopic under areas of mental wellness. The NFPA currently does not include chaplaincy in either NFPA 1500 or NFPA 1583.

The evidence for this book comes from both the fire service research literature as well as research literature on the general public. The main philosophical framework for this book is adopted from Aristotle's teaching on moral responsibility, which is still applied today in the justice system. That framework is presented in chapter 1 and is used to guide the reader in chapters 2, 3, 4, and 5.

Chapter 1 examines and defines the buzzword "resilience." You will learn how resilience is developed, why it is relevant to firefighters, and specifically how your lifestyle habits are linked to your risk factors for heart attack, cancer, and suicide. You will be presented with a summary of the guidance from national fire service stakeholder organizations and where the fire service is with meeting those guidelines. Lastly, you will learn about the philosophical framework that will be applied to chapters 2, 3, 4, and 5.

In chapter 2 you will learn about physical wellness, in chapter 3 you will learn about nutritional wellness, and in chapter 4 you will learn about mental wellness. All three of these chapters follow the framework that you will learn about in chapter 1, which includes three parts: what you cannot change and what you can change, deficiency and excess, and micro and macro. Each chapter provides guidance on navigating physical, nutritional, and mental wellness in a nonprescriptive way. In other words, this book does not tell you specifically how to act. Rather, it educates you on what the circumstances are, what missing the mark looks like, and what factors you should consider. Each one of these chapters finishes with a section that explains the very real consequences of your choices in regards to health outcomes.

Chapter 5 may cause you some stress if you consider yourself a leader. This chapter uses the philosophical concept of necessary and sufficient components to provide an *objective* and assessable definition of the word "leadership." This definition of leadership is philosophically sound and challenges the entire body of literature on leadership both inside the fire service and outside of it. There is a *subjective* nature of leadership, but that should only relate to how you do the *objective* act of leading. In chapter 5, you will be presented with what the components of leadership as an action are.

As the final chapter, chapter 5 ties together the concepts of resilience, health and wellness, and leadership in a seamless way that does not appear in existing leadership literature. It uses the same framework from Aristotle that is found in chapters 2, 3, and 4, but because it is applied to leadership, it does not include the same section of health outcomes at the end of the chapter. Upon finishing chapter 5, you will never be able to look at the topics of resilience, health and wellness, or leadership again in the same way you used to.

> A note about references used in this book: The references used do not reflect an endorsement for any author or establishment. We tried our best to provide logically clear, scientifically accurate, and relevant information. We encourage all readers to have a skeptical eye and, if you are using this book in your fire department, to replace studies or information that is no longer true with the most accurate information you can find.

ACKNOWLEDGMENTS

When I am called to duty, God, whenever flames may rage; Give me the strength to save some life whatever be its age. Help me to embrace a little child before it's too late; Or some older person from the horror of that fate. Enable me to be alert and hear the weakest shout; And quickly and efficiently to put the fire out. I want to fill my calling and give the best in me; To guard my neighbor and protect his property. And if according to Your will I have to lose my life; Please bless with Your protecting hand, my children and my wife.

—A. W. "Smokey" Linn

The first thank you is to The Eternal who gives us life, the free will to choose to be a firefighter, and the free will to choose our behavior.

Thank you to the men and women who serve as career and volunteer firefighters around the U.S. and the world. A special thank you to the firefighters of Howard County, MD, my instructors from HCFR Academy Class 24, my Class 24 classmates, and my shift mates from 9B, SOD-11B, BOSH, 5A, SOD-11A, and SOD. Thank you also those in the U.S. Coast Guard who have mentored and supported me—you know who you are.

Thank you to all the dedicated staff and Advocates of the National Fallen Firefighters Foundation (NFFF) who work around the U.S. (and the world) to honor fallen firefighters, support the families of fallen firefighters, and educate firefighters on how to prevent preventable line-of-duty deaths. Thank you to Rick Best for your mentorship and your tireless efforts to educate the fire service and mentor the next generation of NFFF Advocates. Thank you to Victor Stagnaro and John Tippett for supporting this book as an NFFF project and introducing me to Dr. David Schary. Thank you to David for joining me on this writing journey—I'm proud of what we accomplished and am grateful for the opportunity that brought us together.

Thank you to all the *Fire Engineering* staff who supported this book project. Thank you to Bobby Halton for your friendship and mentorship, for accepting this book for publication, and for the philosophic discussions we shared about it. Thank you for your feedback on the Leadership chapter, which made this book better. See you soon. Thank you to Frank Ricci for your friendship and mentorship. Your feedback for this book made it better. When Bobby introduced us, he said you were one of the bravest men he knew—your words on courage mean a lot. Thank you to Chris Barton and Tony Quinn for your patience and support on making sure that this book was done properly.

Thank you to those in the fire service that played an intimate role in supporting the research and work that was needed to develop this book. Thank you to Joanne Rund for your support and mentorship in health and wellness. Thank you to John Butler for your support and mentorship in my academic pursuits and my military pursuits.

Thank you to those professors who have passed on their knowledge. Thank you to Dr. Christopher Dreisbach for your teaching and mentorship on philosophy, specifically on Aristotle's virtue theory and moral responsibility. Thank you to Dr. Christina Harnett for your teaching on individual and group dynamics and on stress management. Thank you to Dr. Helena Seli, Dr. Monique Datta, and Dr. Eric Canny for your teaching and mentorship on knowledge, motivation, and organizational factors of leadership.

And last but not least, thank you to Dr. Nan Baer, the love of my life. 2023 marked 20 years of marriage, and it keeps getting better. Thank you for your love and support throughout the writing of this book. Gianna, Abram, Moses, and Zohara, I love you all—this book is for you. Thank you Mom and Dad for being positive role models. I owe much of my family success to you. Jay, Danny, Andy, and Joelle, you all are quite hard to keep up with. Keep giving back to the world!

—G

I first want to thank God, who I try to serve each day. Thank you to the wonderful people at the National Fallen Firefighters Foundation who took a chance on an outsider with an idea but no firefighting background—particularly Victor Stagnaro, John Tippet, and JoEllen Kelly. Thank you for the work you do to support firefighters and their families.

Thank you to all the men and women who serve their communities as firefighters. It has been an honor to learn from you. I particularly thank my co-author Gamaliel Baer, whose passion to help his fellow firefighters is inspiring. This book is only possible because of his work. I would also like to thank Kenny

Jenkins for his continued support for firefighter wellness. The Charleston and North Charleston Fire Departments are world-class.

Thank you to the leadership and staff at *Fire Engineering* who believed the ideas in this book were important enough to publish.

Thank you to Alexis Waldren, a friend and colleague, who is solely responsible for introducing me to the firefighting world. Who knew what this would turn into when you invited me to shadow the teaching cadre for the USFS's Human Performance Optimization (HPO) in Sacramento? I appreciate your continued support over the years as I have branched out into other areas of the fire service. Thank you to many HPO cadre members—Josh Eichamer, Ben Iverson, Joe Domitrovich, Kristy Hajny, Charlie Palmer, Joe Sol, Ashley Taylor, and Wanda Wildenberg. I hope one day the band will get back together.

Thank you to the many academic advisors and mentors that took (and continue to take) time to help me grow professionally and personally. I owe my career to Dr. Brad Cardinal, who took a chance on me. I have met few people who are as dedicated, humble, and kind. Thank you to Dr. Don Seigel. You graciously walked a lowly rowing coach through how to do research and write a thesis, then encouraged me to apply to doctoral programs.

Thank you to Brad Wetzel. You told me to write a book. It took me a few years and a prolific co-author, but I finally heeded your advice.

Thank you to all of my colleagues at Winthrop University, and especially the PESH faculty who are led by our outstanding department chair, Dr. Kristi Schoepfer. I could not ask for a better, more supportive boss.

Thank you Mom, Dad, and Dawn. You always stood by whatever decisions I made.

Finally, thank you to the three most important ladies in my life—Lisa, Grace, and Joy—for your continued, unconditional love and support through all of my many endeavors and outlandish ideas.

—D

INTRODUCTION

We wrote this book to address a gap in the fire service literature on health and wellness education. The fire service has made great strides in bringing the issues of heart attack, cancer, and suicide to the forefront of discussion. Those strategic-level ideas exist in the form of guidance and standards from national-level fire service organizations. Strategy is about the end goal or mission. The strategic guidance for the fire service is that health and wellness should be addressed. Tactics are about *how* to do something. There are many ways to be healthy. The tactical choices you will make on how to accomplish your wellness goals will likely be different from other firefighters. Your department may differ from other departments. In between the strategic and tactical levels is the operational level. The operational level addresses the battlefield, or what is known about the different factors at play. Once the different factors are known, all the different possible courses of action can be considered. Operational intelligence explains *what* (and who) is involved.

At the time of publication of this book, no model curriculum existed for fire departments or firefighters that explained the *what*. This book offers the operational intelligence on health, wellness, and leadership. A model fire service curriculum should address the components of the issue at hand and the options and consequences of different behavior, but should not prescribe specific behavior to all firefighters because every firefighter is different. Understanding *what* is involved in health, wellness, and leadership, and not *how* to do each behavior, is the key to operational intelligence. We offer this book to fire departments and firefighters as a comprehensive curriculum that can be adopted for initial or continuing, physical and behavioral, health, wellness, and leadership education.

—G & D

In 2019, Dr. Gamaliel Baer and Dr. David Schary were introduced by the staff of National Fallen Firefighters Foundation (NFFF) because they were each individually working on projects addressing health and wellness for the fire service. Dr. Schary is an associate professor of exercise science at Winthrop University in South Carolina. He had been working with the NFFF on a Human Performance Optimization project for wildland firefighters in the northwest region of the U.S.

Dr. Baer is a professional firefighter for Howard County Fire and Rescue in Maryland. He was the health and wellness coordinator in charge of the peer fitness team, peer support team, and health and wellness education for the fire academy, in-service training, and officer development. He had written his dissertation on health and wellness education in the fire service and he was working on putting the curriculum he developed for the fire academy into a book for the NFFF.

It didn't make sense for two projects that were so similar in concept to move forward without coordination. Gamaliel and David agreed to look at the ways in which they could work together to produce something great for the NFFF and for the fire service as a whole. Because of David's expertise in sports psychology, he wrote the Mental Wellness chapter, and as a highly published university professor, David also provided a thorough review to the entire project. Gamaliel is a certified health coach and certified personal trainer and had already done heavy research into firefighter-specific health and wellness before, during, and after his doctoral dissertation. He had produced and refined an academy curriculum on all the chapters of this book. Based on his education in philosophy from his masters and doctoral education, he had produced a framework using Aristotle's philosophy of moral responsibility that was adopted as the framework for this book. He wrote the Resilience, Physical Wellness, Nutritional Wellness, and Leadership chapters.

Gamaliel and David both feel strongly that the fire service should be providing an adoptable model curriculum for health and wellness, especially since the fire service is making health and wellness such a major focus. Unfortunately, since that was never done, for-profit health and wellness education companies have tried to fill that void and provide curricula and education to first responder agencies at a high cost—up to $1,500 a person. Usually, those costs are covered through grant funding from the Federal Emergency Management Agency (FEMA). Additionally, these companies do not yet offer train-the-trainer programs so that fire departments could develop their own sustainable health and wellness programs. There is no long-term money in that.

A few books in the fire service literature cover health and wellness, but they fall short of being comprehensive enough for fire departments to adopt as an in-house health and wellness curriculum that also educates on heart attack, cancer, and suicide. We hope that, eventually, the focus will shift toward helping fire departments build sustainable in-house health and wellness programs that include heart attack, cancer, and suicide risk reduction education programs instead of building a fire-service-industrial-complex that focuses on making money through third party services.

The NFFF serves to honor fallen firefighters, honor and serve the families of those fallen firefighters, and educate the fire service on how to prevent line-of-duty deaths. As a 501(c)(3), the NFFF is not in the business of making money and is solely focused on the good of the fire service. The NFFF was founded by Congress, and continues to be grant-funded so that it can offer education to the fire service at little or no cost to fire departments. The NFFF, as well as both Gamaliel and David, assert that this book can serve as the national curriculum model and as a primary source for educating fire departments on health, wellness, and leadership. As per NFPA 1500 and NFPA 1583, health and wellness education for firefighters should be a discipline that can be developed and maintained in any fire department that has the desire to do so. This should be possible without having to pay hundreds of thousands of dollars a year to a for-profit company.

To that end, Gamaliel and David have committed 100% of their book royalties to the NFFF. They offer this four-year book project as the first comprehensive fire service health, wellness, and leadership education curriculum that was produced by the fire service and for the fire service. Gamaliel, David, and the NFFF hope that this can be the beginning of fire service scholarship on a nationally adoptable health and wellness curriculum model, but not the end of it.

1
RESILIENCE

Firefighters are some of the most cherished members of the community. They are heroes, putting their lives on the line in order to save people, property, and the environment. The job can be dangerous—even fatal—but firefighters willingly accept the risks to serve the public. Unfortunately, firefighters are more likely to die from lifestyle-related heart attacks than from any other job-related death.[1]

No one should ever have to question whether the death of a firefighter was due to an unhealthy lifestyle. The community expects firefighters to be physically fit, mentally fit, and ready to respond to any emergency or rescue anyone in need. Unfortunately, not all firefighters meet that expectation.[2]

There are organizations around the United States addressing the health of firefighters, specifically trying to prevent heart disease, cancer, and suicide. For example, some organizations have advocated for firefighter deaths from cancer or suicide to be treated as job-related deaths (heart attacks are already treated this way). Fire departments and firefighters, however, must demonstrate that they are doing everything they can to stay healthy. This will ensure that fatalities are not the result of poor mental or physical health.

Firefighters are passionate about their job, as well as protecting and serving their community. Firefighters may naturally have more interest in one specialty over another, like enjoying search and rescue more than emergency medical services (EMS), but all firefighters should share one thing as a common passion—health and wellness.

Why is Health and Wellness Important?

We are what we repeatedly do. Excellence, then, is not an act, but a habit.

—Will Durant

Durant's quote, which paraphrases Aristotle's teachings, sets the foundation for how you should think about health. To help us understand how this works, let's break this quote down into two parts:

1. *We are what we repeatedly do.* This phrase describes that, as people, our present condition is the result of our actions.
2. *Excellence, then, is not an act, but a habit.* We don't become excellent from one act; rather, it is from repeatedly choosing the right behaviors over time.

Firefighters should consider their health through the lens of Aristotle's teaching. Health is not decided from one decision or action. It comes from a pattern of behavior repeated over time. We could easily rewrite Durant's quote to read, "We are what we repeatedly do. Health, then, is not an act, but a habit."

Defining Health and Wellness

Health is a snapshot of your body, mind, and soul, which is the result of your wellness efforts.

Health and wellness are not the same things. Merriam-Webster defines health as being sound in body, mind, or spirit.[3] When you go to your primary care physician for a checkup, your physician will record metrics of your physical body at that moment in time. These metrics are used to determine if you are within a normal range or whether there is an indication of disease. Humans of all backgrounds, sizes, shapes, and conditions of physical fitness can have health metrics that fall within a normal range. Similarly, a mental health professional or spiritual counselor can make determinations as to whether a person is within a normal range or if there are abnormalities that may need to be addressed.

Wellness refers to the efforts you make to improve or maintain your health. Wellness efforts might include an exercise routine (physical activity); adherence to certain dietary standards (nutrition); habits that increase your knowledge, motivation, concentration, or mindfulness (psychology); and habits of love, forgiveness, and connection (spirituality). In other words, wellness is what determines the health of your component parts as a human. Wellness is the input, and health is the outcome.

The idea that health is brought about by wellness habits may sound too simple. It is true that some elements of life cannot be controlled, and in some instances no amount of wellness efforts will change a person's health outcome. Some tragedies occur due to genetic mutations or exposure to toxins. However, your wellness efforts play a major role in your health, and as we hope to demonstrate in this book, the control you have over your health outcomes is worth knowing.

Defining Resilience

Resilience has become a buzzword in the health and wellness industry, as well as in the fire service. Resilience is related to health and wellness, but has its own specific meaning as well. Merriam Webster offers two definitions for resilience.[4] The first pertains to the science of physics and defines resilience as the capability of a strained body to recover its size and shape. The second definition is more applicable to humans and defines resilience as the ability to recover from or adjust easily to misfortune and change. Perhaps because the word resilience is used for both physics and humans, different uses of the word have been adopted. Three common uses of resilience include the following:

1. The ability to bounce back[5]
2. Having more in the tank (i.e., reserves)[6]
3. The ability to overcome a challenge[7]

Bouncing Back

When we talk about resilience in terms of humans, it becomes clear that the first common use of the word is incomplete. Firstly, humans naturally change with age, but they can also change through wellness efforts. Steel might also change with age over time but cannot become more resilient like humans through wellness efforts. But "bouncing back" as a definition of resilience must be understood as a snapshot in time—like the definition of health—that can change since humans are dynamic, as opposed to steel which is relatively static.

Secondly, humans are not one-dimensional. Humans have a body, mind, and soul (or bio-psycho-social dimensions, if you prefer), and thus bouncing back would have to relate to more than just your physical body. Yes, you can bounce back if you push on your skin or muscles, and you can also bounce back if you get sick or injured. However, humans have a mind and a soul as well, and so the idea of bouncing back would have to account for those dimensions as well. Two individuals could have the same physical resilience but different mental or spiritual resilience, and that could determine if someone bounces back or not.

Lastly, if bouncing back is the only aspect of resilience, then the saying, "If you get knocked down seven times, stand up eight," makes sense. But if you are in a boxing ring and Mike Tyson is the one knocking you down, then continuing to stand up may eventually lead to your death! Calling a person resilient in a situation that may lead to death, when there are better options to help you survive, misses the point. In fact, another saying may be more appropriate for that example: "Insanity is doing the same thing over and over and expecting a different result."

Having Reserves

The second common use of resilience is connected to the first. Having reserves is about preparing for a challenge. Your reserves, however, are context specific. If you get knocked down, it is possible that you do not get up again, especially if you did not have the physical reserves necessary to take the blow. An NFL linebacker should be able to take a physical hit better than a long-distance runner. While a long-distance runner may not be able to take a physical hit compared to the NFL linebacker, they are able to maintain an elevated heart rate for a longer period of time. The context of the challenge matters.

Although reserves are necessary in order to bounce back from a challenge, resilience cannot just be only about reserves. If reserves are what make humans resilient, then more is simply better. In other words, being as muscular as possible, as fat as possible, or as rich as possible should make you more resilient because you would always have more reserves. It is possible that those specific reserves could help someone be resilient to a specific challenge, similar to the NFL linebacker and the long-distance runner. However, some of the most resilient people in the fire service or the military are not the strongest, fattest, or richest people. We do not call people resilient just because they have a lot of a specific reserve. There seems to be something else.

Overcoming a Challenge

The third common use of resilience includes both the first and second uses, but adds a qualifier. If you get knocked down, and you get back up again, you have demonstrated that you bounced back, probably because you had enough reserves. However, if you get knocked down seven times and get back up eight only to be knocked down again in the same exact way, then you have not overcome the challenge.

Using Merriam Webster's second definition, you may have "recovered from" but not "adjust[ed] easily to misfortune and change." You do not want to simply get back up after being knocked down by the same challenge; you want to stop getting knocked down and overcome the challenge (if possible). Humans are not static like steel. You can adapt and overcome greater challenges than you did in the past. But to do so, you need to grow your abilities and change for the better.

Resilience is a measurement of how much challenge your body, mind, and soul can overcome. Figure 1–1 offers a theoretical equation for resilience. It states that resilience is the product of your body, mind, and soul divided by the challenge you are faced with. If you have abilities that are at least equal to the amount the challenge requires, then you will survive, but you may need to recover. If you have more ability than what the challenge requires, then you can defeat the challenge with some reserves left over.

Any two people will have different baseline levels of resilience, because each person is genetically different. However, if you compare two people with the same baselines, resilience is determined by conditioning, or the effort put in to grow their abilities. All things being equal, the person who has better physical conditioning will be able to overcome more physical challenges. Similarly, with the same IQ as a baseline, the person who learns and practices a skill more will be able to perform better than the person who does not. For those that believe love and forgiveness are abilities of the soul, the one who practices loving and forgiving others is better conditioned in those areas.

Ultimately, you should only compare yourself to yourself, not to other people. You cannot change your genetics, but you can work to optimize your health and abilities. The more you improve, the more challenges you can overcome.

$$R = \frac{B \times M \times S}{\text{Challenge}}$$

FIGURE 1–1. An equation for calculating resilience

Growing Human Resilience

As a human, you are unlike steel in that you can change for better or worse. You can become more (or less) resilient as it relates to the ability of your body, mind, or soul to overcome a challenge. That ability might be specific or it might be general. But in either case, in order to be more resilient next year than you are right now, you have to be able to overcome a harder challenge. In other words, you have to grow in one or more of your human dimensions (body, mind, or soul).

Building resilience in your body, mind, or soul takes time. As a firefighter, you should build a certain amount of resilience to prepare for specific challenges you might expect to face during a shift. Improving your ability to do specific actions related to firefighting, like forcible entry or dragging a downed firefighter, improves your specific resilience to a challenge. But firefighters should also build a certain amount of resilience to prepare for challenges that are more general in nature. For example, increasing your aerobic fitness will give you the ability to work longer, which means your air cylinder will last longer, allowing you to do more when responding to an emergency. Humans are adaptable and can grow the ability to deal with specific or general challenges.

In order to grow more resilient, you must go beyond your normal range or your habitual limits (fig. 1–2). Injury occurs when you face a challenge to your body, mind, or soul that you were not prepared to overcome. There may be theoretical limits to what you can do with your body, mind, and soul, but each person is different. No matter how resilient you grow in any of the human dimensions, you may eventually be faced with something you cannot overcome alone. How do you overcome a challenge that is beyond your own ability?

If resilience were only the ability to bounce back based on your reserves, then whenever you are faced with a challenge that is beyond your ability, you

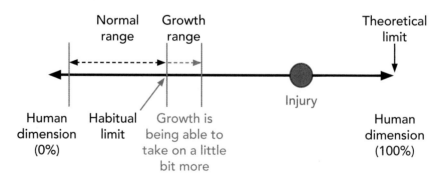

FIGURE 1–2. Growing human resilience

would fail. For physical challenges where you do not have time to seek support, this could be a life-and-death issue. However, many other types of challenges in life are not physical, but rather mental or spiritual challenges.

Fortunately, humans can benefit from networks of support to overcome challenges. Someone else has likely faced the same challenge you are dealing with. As a firefighter, you often turn to a peer to learn how to overcome a challenge. This is one of the secrets to being a more resilient person. Learning from others can help you to overcome challenges that you could not overcome on your own. Therefore, fire departments would benefit by investing in their ability to provide continual health and wellness education to their members.

A Need for Health and Wellness Education

Resilience improves as you increase your ability to overcome challenges. In order to increase your abilities, you must make an effort to have better wellness habits. If wellness habits lead to better health and abilities and thus greater resilience, then wellness-related education should be a core part of the fire service and health care systems in general. There is evidence, however, that this is not the case for health care systems in the United States. For example, Dr. Bradley Cardinal looked at the curriculum for 118 medical doctorate programs around the United States and found that education about the health benefits of exercise were rarely required, if ever.[8] The study concluded that over half of U.S. physicians that graduated in 2013 received no formal education on the health benefits of physical activity. This lack of exposure could affect how they educate and treat patients. While we cannot blame the country's health issues on physicians or the health care system alone, Dr. Cardinal's findings highlight that the American health care system prioritizes treating the symptoms of disease instead of preventing the disease. Prevention is harder because it requires behavior change, but it is less expensive and more effective.

Inspired by Dr. Cardinal's study, Dr. Baer analyzed the published curricula from state-sponsored fire service training programs from eight states in the Mid-Atlantic region (VA, MD, DE, PA, NY, NJ, MA, and CT).[9] He also examined three additional states that represented the political spectrum of liberal, moderate, and conservative (CA, OH, and TX, respectively). The analysis focused on training that would address health and wellness for firefighters with an effort to prevent heart attack, cancer, or suicide. At the time of the study, only two states (MA and NJ) had classes available to firefighters that addressed health and wellness (MA had one and NJ had three). The remaining nine publicly available state training curricula contained no health and wellness courses available to firefighters. These states employed 110,200 career firefighters, roughly a third of the 315,910 career firefighters in the United States.

Does your state offer courses to firefighters on health and wellness? Does your academy? Does your department? Three areas that a fire department should focus on if it commits to educating on health and wellness are the Big Three (heart attack, cancer, and suicide); lifestyle factors and the role they play in health outcomes; and the national-level guidance and law that exists for the fire service on health and wellness.

The Big Three: Heart Attack, Cancer, and Suicide

Three types of fatalities that are tracked in the fire service are heart attack, cancer, and suicide. These fatalities are not only affecting firefighters, but also the U.S. general population at large. According to the Centers for Disease Control and Prevention (CDC), the leading cause of death for all ages is heart disease, followed closely by cancer.[10] Suicide is the twelfth leading cause of death when considering all ages. However, the age range is a key issue.

Firefighting as an occupation does not include all ages, as represented in national fatality statistics. When you constrict the fatality statistics to the age groups that most closely resemble career firefighters (ages 15–65 years old is as close as the National Institutes of Health data allows), the leading causes of death for each age group change substantially. Suicide jumps from twelfth for all ages to fifth for ages 15–65 (prior to the COVID-19 pandemic, suicide was fourth in the 15–65 age group). Cancer becomes first and heart disease becomes second. Besides cancer and heart disease, only COVID-19 and unintended injury outnumber suicide (and there continues to be controversy around how COVID-19 deaths were counted).

Table 1–1 shows the raw numbers of the U.S. general population from 15–65 years old who died from the top five causes of death for that age group (at the time of publication, the most recent fatality data from the CDC was for 2020). This information is significant because it means that heart attack, cancer, and suicide are major issues for the entire population of the U.S. and that firefighters are not the only group facing these issues. Firefighters certainly deal with more stress and are exposed to more trauma than the average person. Yet, firefighters and researchers studying the Big Three in the fire service must be cautious not to overstate the problem as something firefighter-specific. It is clear that these issues exist in both the general population and the fire service. This could be because the lifestyles of the general population and firefighters are very similar.

TABLE 1-1. Top Five Causes of Death for U.S. General Population (2020)

Rank	Cause of death for ages 15–65	Total number for ages 15–64
1	Cancer	160,441
2	Heart disease	139,751
3	Unintended injury	134,223
4	COVID-19	67,888
5	Suicide	36,239

Lifestyle and Fatal Health Outcomes

Heart attack, cancer, and suicide often occur as a result of physical or behavioral health issues. It is possible to have both physical and behavioral health issues at the same time. The research shows a connection between these three health issues and lifestyle. Baer considered these connections by analyzing research on both firefighters and the general public.[11] A graphical representation between health issues and lifestyle factors was created to more easily see the connections being considered. In figure 1–3, each line represents at least one research article that established the connection being depicted.

The research shows that heart attack, cancer, and suicide can be the fatal result of poor physical and behavioral health. While genetics can play a role in tragic health outcomes, your health is more often the result of your wellness habits, specifically physical activity, nutrition, and mental activity. While figure 1–3 is not a comprehensive list of all physical and behavioral health

FIGURE 1-3. Connections between lifestyle and the Big Three

issues, the message is clear. Your wellness habits influence your physical and behavioral health, and poor physical and behavioral health issues that are ignored or unaddressed increase the likelihood of the Big Three fatalities.

National Guidance, Standards, and Laws

According to three of the highest-profile fire service organizations in the U.S. (National Fallen Firefighters Foundation [NFFF], International Association of Fire Fighters and International Association of Fire Chiefs Wellness-Fitness Initiative [IAFF/IAFC-WFI], and National Fire Protection Association [NFPA]), fire departments should be working to improve the health and wellness of their firefighters. The fire service cannot in good faith continue to ignore the science on physical and behavioral health. There have been major changes and advances within the fire service, and advocacy for health and wellness continues to build. More importantly, major stakeholders of the fire service industry offer guidance, standards, and even laws about the issues of health and wellness, the importance of researching those issues, and the importance of educating firefighters about them. Figure 1–4 lists the most pertinent guidance, standards, and laws applicable to the fire service.

Firefighters should know what guidance, standards, and laws exist that serve to protect their interests. The Occupational Safety and Health Administration (OSHA) has regulations that employers must train their employees on, including how to reduce or eliminate work-related illness, injury, and death. The Public Safety Officers' Benefit Act (PSOB) is a federal law that allocates tax dollars for families of fallen public safety officers who have died in the line of duty. Currently, the PSOB includes coverage for certain heart attacks and

	Education	Wellness programs	Research
NFFF	Life Safety Initiatives #1, #5	Life Safety Initiatives #6, #13	Life Safety Initiative #7
IAFF/IAFC-WFI	WFI 4th edition, p. 8 (mission statement)	WFI 4th edition Ch. 3, 5	WFI 4th edition Ch. 7
NFPA	1500 Ch. 5 1583 Ch. 7, 8	1500 Ch. 4, 11, 12, 13 1583 Ch. 5, 6, 7, 8	1500 Ch. 4, 11 1583 Ch. 9
OSHA	Training requirements, pp. 1–4		
NIOSH		Kunadharaju, Smith, and Dejoy (2011) Hard et al. (2018)	

FIGURE 1–4. Summary of guidance, standards, and laws that affect the fire service

strokes, which can each be the direct result of lifestyle factors. Interestingly, 90% of firefighters who die of a heart attack on duty have underlying cardiovascular disease.[12] Additionally, even though the number of overall firefighter line-of-duty deaths (LODDs) has declined, the large majority of firefighter LODDs continue to be cardiac related. OSHA's education manual states that employees should not have to become ill, sustain injuries, or die for a paycheck, and that it is the responsibility of the employer to educate employees on how to prevent work-related illness, injury, and death.[13] State and municipal governments that regulate fire departments must meet or exceed OSHA's training regulation. Additionally, the NFPA has included that regulation in its standards, including chapter 4 of *NFPA 1500: Standard on Fire Department Occupational Safety, Health, and Wellness Program* and chapter 8 of *NFPA 1583: Standard on Health-Related Fitness Programs for Fire Department Members*.

> NFPA 1500: chapter 11 covers medical and physical requirements; chapter 12 covers behavioral health and wellness programs. NFPA 1583: chapter 6 covers fitness assessments; chapter 7 covers exercise and fitness training programs; chapter 8 covers health promotion education.

Individual and Organizational Responsibility

Fire departments have a responsibility to educate and train their firefighters. For example, the fire department is responsible for providing the appropriate personal protective equipment (PPE) and training the firefighters on proper use of that equipment. After being equipped and trained, the firefighter has a responsibility to properly use the equipment when engaging in work-related activities requiring PPE. If during an emergency a firefighter fails to use the proper PPE or uses PPE the wrong way, an investigation is likely to take place to determine who was responsible for the error. Thus, if the argument from the fire service is that heart attack, cancer, and suicide are job-related, then fire departments should provide appropriate education on how to prevent those issues. At the very least, firefighters should be taught how to prevent heart attacks and strokes since both are federally covered as LODDs in the PSOB.

Determining responsibility can be difficult. Fortunately, Aristotle teaches a method to accurately assess responsibility. Figure 1–5 shows Aristotle's four steps to assess responsibility, which can be applied to both the individual and organization.

Did the Individual Know the Circumstances?

Aristotle taught that there are things that cannot be changed and things that can be changed. It is critical that individuals and organizations know the difference when assessing a situation. Understanding the things that cannot be changed may be important for planning and overcoming challenges. However, the effort that individuals and organizations make in any endeavor should be put into things that can be changed. To navigate the things that can be changed, there has to be knowledge of right and wrong.

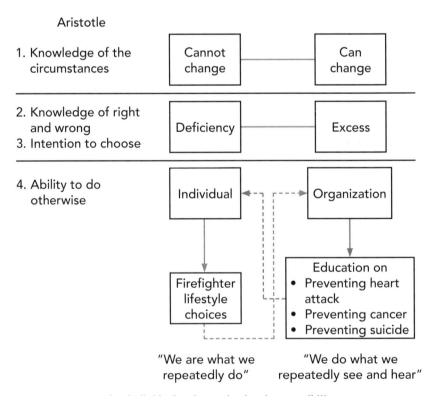

FIGURE 1–5. Assessing individual and organizational responsibility

Did the Individual Know Right From Wrong?
Aristotle taught that virtue (right) is the mean between the extremes of deficiency and excess (both being wrong). Deficiency is a lack of something, and excess is too much of it. An example that Aristotle gives is courage. Courage is the mean between cowardice (deficiency) and foolhardiness (excess). When it comes to health-related matters like eating, Aristotle said that the balance is relative to the individual. Deficiency and excess of caloric intake are different depending on someone's size and activity level. But knowing right from wrong does not guarantee an individual or organization will make the right choice.

Did the Individual Voluntarily Choose Their Action?
Aristotle taught that the individual must know the difference between deficiency and excess for the practical issue at hand, and then the individual must voluntarily navigate the options available in a way that is balanced. That does not mean you have to constantly remain at some arbitrary 50% level of whatever you are doing. For instance, you cannot eat consistently in order to maintain 50% fullness. There is a spectrum of deficiency (not eating) and excess (eating until you are full), and it is healthy to engage in both. The key is to attempt to find the balance of engagement. It is possible to have the intention to choose a balanced path, but be prevented in carrying out that choice.

Did the Individual Have the Ability to Do Otherwise?
Aristotle taught that an individual's choice can be influenced so that a person's ability to choose is limited or even restricted to only one choice. This can be seen in differences of power between two people, like a firefighter and the shift officer, or a child and the parent. It can also be seen on a larger scale between an individual and the rules of society, or even the individual and the environment. Interestingly, this can also apply within a person, specifically in terms of the way a person chooses to live and the way their body's systems, organs, and cells are forced to react to those choices. All of these are examples of a relationship between the smaller (micro) and the larger (macro). It is important to remember that individuals make up an organization just like cells, organs, and systems make up an individual. This means that, for organizations and for humans, there is a feedback loop because the macro is made up of the micro. The relationship between the micro and macro is critical to understanding health optimization, because your body systems, organs, and cells cannot choose

for themselves. They are forced to deal with the choices that you make as the chief health officer of your own body.

If any of the answers to Aristotle's four assessments are "no," then the individual cannot be held fully responsible (for better or for worse) and an investigation is required. The same assessments can be applied to an organization to determine its responsibility. Assessing both individual and organizational responsibility is important because, according to chapter 4 of NFPA 1500 and chapter 8 of NFPA 1583, fire departments should be providing training to reduce job-related illness, injury, and death. Firefighters are responsible for applying the training they receive in the academy and throughout their careers. However, fire chiefs must understand that their departments are made up of individuals, so what firefighters are taught will eventually be reflected by the organization's performance. We are what we repeatedly do. However, we do what we repeatedly see and hear.

The Change Assessment Model

The image in figure 1–5 can be cumbersome, so a simplified version will be used throughout the book. Figure 1–6 graphically presents Aristotle's four assessments using three pairings. Nature and nurture represent the circumstances of what you cannot change (nature) and what you can change (nurture).

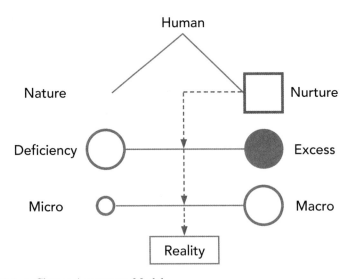

FIGURE 1–6. Change Assessment Model

The spectrum of deficiency and excess exists for both knowledge of right and wrong as well as choice. Assessing the relationship between micro and macro allows you to identify the forces that may influence your choice. Your reality is determined by the impact of all of these forces on you and how you navigate them.

Over the course of this book, this Change Assessment Model will be used as a way to organize the chapters on physical wellness, nutritional wellness, mental wellness, and leadership. The Change Assessment Model is generalizable and can be used as a template to design your own curriculum entirely or modify these chapters in order to fit your personal needs or your department's needs. The adoption of this model is encouraged to organize health and wellness courses in your fire department in a logical way that can be compared, assessed, and improved upon by other departments or your own department over time.

Nature and Nurture

Despite your daydreams at times, firefighters are humans too. The laws of physics, biology, and chemistry apply to your body. In regards to fitness, you have to understand the circumstances of what cannot be changed and what can be changed. Nature and nurture are words that are used in this book to describe those two ideas respectively. When considering health and wellness issues, it is important to understand that to some degree your health is determined by your genetics. But even the human genome, your genetic makeup, is still a product of nature and nurture.

In 1990, the Human Genome Project began as an international collaboration to map the entire human genome.[14] The project was completed in 2003. Prior to the project, estimates were made as to how many genes existed in the human genome, some of which exceeded 100,000. As it turned out, there are roughly 30,000 genes in the human genome.[15] This number puzzled researchers, because 30,000 genes did not seem high enough to account for all of the diversity within the human population. Eventually, researchers discovered that each gene can be expressed in different ways, known as being either up-regulated or down-regulated.[16] In other words, each gene is like a switch, and with 30,000 switches that can either be up or down, researchers realized there is plenty of opportunity for diversity. Epigenetics is the study of gene expression, examining the factors that cause the switches to be either up or down.

You inherit a genetic code from your parents. Your genetic code provides information about the size of your skeleton and the makeup of your muscles. It provides information about the color of your skin, hair, and eyes. It provides information about your brain structure as well as other organs. Your genetics determine many things about you that you cannot easily change. In other words, you have physical traits that make you who you are—this is your *nature* component. However, thanks to the Human Genome Project, we also know that there are ways to change your genetic expression by up-regulating or down-regulating genes. The choices you make and your environmental exposures are the main factors that affect your gene expression—this is your *nurture* component. It is important for firefighters to understand that your lifestyle choices play a major role in your health outcomes—they can literally change your genes.

Understanding what you cannot change about your body gives you a better understanding of the circumstances that you face in your efforts to be well. Aristotle taught that the first area of importance when analyzing choices and responsibility is whether the individual understands the circumstances. The *Nature and Nurture* sections of each chapter will present circumstances that can and cannot be changed.

Deficiency and Excess

Recall that virtue is the mean between two extremes. In order to find a balance, you have to know what the extremes are. The *Deficiency and Excess* sections present information on the extremes of physical wellness, nutritional wellness, mental wellness, and leadership; these extremes are things you can change, and these sections will provide guidance on how to balance those extremes. Relativity is the dependence of one thing on another. For practical matters like physical wellness, nutritional wellness, and mental wellness, the right amount is relative to your needs. Some people will need more sleep than others, some people will need more protein than others, and some people will need more time alone than others. Finding a virtuous middle does not mean avoiding extremes of activity such as sleep or sprinting, drinking only water or eating a big meal, or being alone or partying. It means that you are balanced in your approach across the spectrum of all activity. When you are out of balance, you increase the likelihood of developing stress-related health issues that can lead to the Big Three.

In the United States, firefighting is the second most stressful job, only surpassed by enlisted military.[17] Figure 1–7 provides a graph used to represent relative differences between the general population and firefighters in the U.S. The stress of the general population is shown in between 0% and 100%, representing the average person's stress level. Firefighters are shown to be slightly closer to 100% compared to the general population. Right below the stress spectrum is a spectrum that reflects wellness efforts. If it is true that firefighters are more stressed than the average worker, then in order to work toward the virtuous middle, firefighters should put more effort into wellness than the

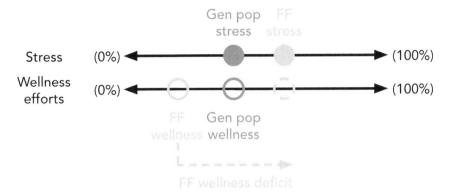

FIGURE 1–7. Wellness deficit of firefighters

average worker. However, the research suggests that not only are firefighters more stressed, a significant number have worse wellness habits than the average worker.[18] Thus, the typical U.S. firefighter faces a wellness deficit compared to the average worker, which might explain why firefighters die more from heart attacks on duty than all other fire service work fatalities.

Even though virtue (the right way) is found somewhere in the middle of two extremes, you are not required to constantly remain at some arbitrary 50% level of whatever you are doing. In fact, in some cases that is impossible. For instance, there are times when you must sleep and times when you cannot eat. There are also times when you are alone by yourself and times when you must work with a group. In order to survive (and thrive) you must be able to handle all aspects of the choice spectrums. In the *Deficiency and Excess* sections, you will learn about those spectrums and why engaging in different parts of each spectrum will help you to create an overall balance of wellness.

Micro and Macro

The last issue regarding responsibility that Aristotle focused on relates to your ability to "do otherwise." In other words, you are only fully responsible if your choices were not limited in some way. Aristotle gave examples of how forces that are larger than you can influence your decisions, such as your shift, station, and department (macro).[19] Likewise, your lifestyle choices influence your body, such as your systems, organs, and cells (micro). The difference is that your body's systems, organs, and cells have no choice but to react based on your lifestyle choices, whereas you do not always have to act based on influences from your shift, organization, or even society. Figure 1–8 is a visual on where you as the human stand in relation to smaller influences (micro) and larger influences (macro).

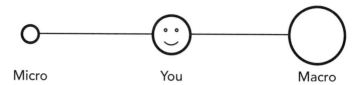

FIGURE 1-8. The spectrum from micro to macro, including the human

A cyclical relationship exists between the micro and the macro. Just as your shift can influence the way you behave, your behavior feeds back into the shift. As a result, the culture of each shift is the product of all the individuals of that shift. Similarly, your systems, organs, and cells are not just influenced by your lifestyle choices; they also shape you as a person.

Therefore, when your systems, organs, and cells are not functioning properly, they can and will affect your decisions, and yet they are just providing feedback based off your decisions. It is true that some people are born with genetic disorders or are affected by an accidental exposure to environmental toxins which may have nothing to do with a person's lifestyle choices. However, as you will see in the *Micro and Macro* sections of this book, lifestyle choices, which you have a lot of control over, play a major role in your health.

This last section begins with a look at how your choices affect your body. As you will find out, heart attack, cancer, and suicide happen to be highly preventable tragedies where lifestyle plays a major role. The relationship between you and the cells in your body will then be expanded on to understand the relationship between your organization and you. You will also read about what forces exist that may be negatively influencing your decisions to become and stay physically and mentally healthy, so that you can identify them when they arise, and seek out support to deal with or overcome them.

Micro: You and The Cell

Your body began as one cell, and that cell divided over and over again until you ended up with the trillions of cells that make your body. Each cell contains DNA, a genetic instruction guide allowing it to contribute to an organ, which contributes to a system, which allows your body to function as a whole. What you choose to do with your body will influence each system in your body, which influences your organs, which influences your cells. There is a constant, two-way feedback loop between the cell (micro) and you (macro), as shown in figure 1-9.

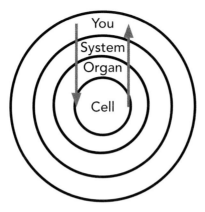

FIGURE 1–9. Your relationship with your systems, organs, and cells

Physical wellness, mental wellness, and nutritional wellness all affect the systems, organs, and cells in your body. There is mounting research on how each of these activities plays a role in cardiovascular, cellular, and neurological health. In order to reduce your risk of the Big Three, you must take control of your lifestyle.

Heart Attack

Since the fire service began recording LODDs, the number one killer has been heart attacks, averaging roughly 50% of all LODDs. Firefighting is a stressful job, and firefighters are at times pushed to their physical limits. The reason for heart attacks in the fire service, however, cannot be solely attributed to the physical stress of the job. In fact, the large majority of firefighters who die of a heart attack on duty are dying due to preventable lifestyle-related issues. As already mentioned, of all heart attack related LODDs, 90% occurred in firefighters that had cardiovascular disease (CVD). Lifestyle issues like hypertension, obesity, unhealthy cholesterol profiles, and diabetes mellitus are implicated as contributing factors.

Cancer

Cancer is another major health issue that affects both the general population and firefighters. It is the second leading cause of death for the general population after heart disease. You should be aware that cancer turns out to be highly preventable too. An expert team of researchers from the MD Anderson Cancer Center at the University of Texas-Houston found that 90%–95% of all cancers are from lifestyle or environmental factors, not genetics.[20] Separate studies have found that 44% of cancers can be attributed to preventable risk factors.[21] However, unlike the study out of MD Anderson Cancer Center, the other studies did not compare prevalence of purely genetic cancers with lifestyle or environmental cancers. You should also know that as a firefighter you are only slightly more likely to get cancer than the general population (9% higher), and only slightly more likely to die if you get cancer (14% higher).[22] It is not clear, though, how much of the incidence rate or mortality rate increases might be from lifestyle factors. Tobacco, poor diet, and lack of exercise were noted as common factors for firefighters.

Suicide

Suicide is the twelfth leading cause of death for all age groups in the general population. But as mentioned already, it jumps to the fifth leading cause of death for working-age Americans. You should be aware that in-depth investigational studies into suicide, called psychological autopsies, have repeatedly shown that 90% of people who die by suicide have diagnosable, and thus treatable, mental disorders.[23] As it turns out, lifestyle has a profound impact on reducing some major mental disorders:

- Depression[24]
- Anxiety[25]
- Stress[26]
- ADD/ADHD[27]
- Addiction[28]

All of the mental disorders listed are contributing factors to suicide.[29]

Interestingly, a 90% preventability rate is a consistent theme across heart attack, cancer, and suicide. You can greatly reduce your risk of heart attack, cancer, and mental disorders by optimizing your wellness behaviors. Research continues to support that physical wellness, mental wellness, and nutritional wellness all influence your body's systems, organs, and cells.

Knowing that something is good for you does not necessarily translate into doing it. For example, in one study, researchers found that almost everyone (99.6%) agreed that exercise is good for you.[30] However, many of the study participants lacked specific knowledge about which diseases are associated with a lack of exercise. Those who knew more about how exercise reduced the risk for specific diseases exercised more often. Therefore, fire departments must go beyond simply telling firefighters that a certain behavior is good for you and do a better job of explaining the consequences of your behaviors.

Macro: The Group and You

As Aristotle taught, your choices can be affected by forces that are larger than you. Yet for the most part, the forces that act on individuals tend to overlap regardless of whether the activity is physical wellness, nutritional wellness, or mental wellness. Therefore, some of the common macro forces that you may be faced with throughout your career are presented in this section to avoid repeating the same themes in each chapter.

People do not make their choices in a vacuum. Aristotle was aware of this, which is why he wanted to know whether someone had the ability to act differently than they did before he determined whether or not they were to be praised or blamed for their action. Therefore, in order to fully consider the actions that you can take for your wellness efforts, you must be aware of the forces that influence you so that you can try your best to avoid negative influences whenever possible.

You, as a whole person, are in a macro position to your cells, but you are in a micro position to a group. In the fire service, there are many levels of influence above the firefighter position. Those levels extend from the company to the department level, and even into local government. Figure 1–10 shows the relationship of the individual firefighter and the fire department.

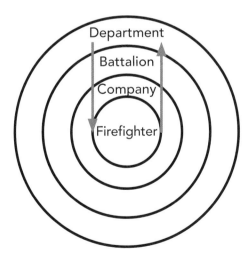

FIGURE 1–10. The department's relationship with the battalion, company, and firefighter

It is important to know that even just one person who has your attention can influence you. You can be influenced by the music you listen to, the art you look at, the movies you watch, or the literature you read. Albert Bandura, a famous social psychologist from Stanford University, has produced some of the leading literature on how humans learn from copying others' behavior. Bandura's research went on to be used in the entertainment industry, where actors model behavior that producers want viewers to imitate.[31] The idea is known as "entertainment education." Keep this idea in the back of your mind then next time you read an opinion piece in an industry magazine or tune in to your new favorite podcast. What others want you to think and do is not always healthy.

Most relationships operate in a reciprocal manner. You affect other people and they affect you. Each interaction changes you and the environment at large. A family or organization's culture is the product of continuous interactions between people. This is evident within the fire service. The culture of the fire service shapes the firefighters and vice versa.

Knowing that your actions can be influenced by your family, work, or even pop culture will hopefully empower you to identify situations that could cause harmful habits. You cannot always avoid a situation, which is why Aristotle said you must assess a person's ability to do otherwise. In those cases, you have to do your best within the situation. Three organizational factors that can influence your wellness habits are culture, rules, and resources.

Culture

Culture refers to the ideals, norms, and priorities of a group of people. For firefighters, a certain culture can exist for an entire department, while separate cultures can exist between shifts, battalions, and even stations. If you are assigned to a station that has a culture of sedentary behavior, over time you may feel that the behavior of the group is normal and may eventually become more sedentary yourself. It is hard to behave in a way that goes against the norm. In fact, health questionnaires that ask about family history are not just looking for genetic clues—they are also looking for clues of group behavior.[32]

In 2001, and again in 2012, data on the firefighter fitness culture found that most overweight firefighters did not believe they were overweight.[33] In 2008, research found that if firefighters could do the job, they were considered fit, even if their health metrics were considered unhealthy.[34] Another research study found that several factors influenced firefighter health, including station eating habits, supervisor leadership and physical fitness, and age and generational influences.[35] A consistent theme across all of these research studies was that coworkers and supervisors have a direct effect on the perceptions and health behaviors of firefighters.

The effect of others operates on a biological level. Within your brain, mirror neurons and limbic resonance are two biological mechanisms that can account for an organization's culture. As discussed in more detail in chapter 4 on mental wellness, mirror neurons are brain cells that detect the emotions or experiences of another person and mirror them within you. In essence, they translate outward behaviors to inner states. Seeing someone who is crying is likely to produce feelings of sadness because you're translating the outward sign of crying as the inward state of being sad.

Within families, friendship groups, and organizations, you observe other people and they observe you. Observing another person, without being distracted by your own thoughts or feelings, allows you to empathize with that person. Your brain mimics the other person's feelings. To successfully navigate this, you need to be mindful of your own behavior and thoughts. Your ability to navigate this can help you stay present and better relate to other people instead of either focusing on yourself or getting completely hypnotized by other people's behavior.

Limbic resonance describes the activity in your brain that occurs when you share a meaningful connection with someone. This happens because your limbic systems actually connect with each other. Being seen and heard by another person helps regulate your brain activity. As a result, you develop stronger connections with that person. When scaled to the company level, station level, or entire fire service organization, this has a powerful effect on workplace culture and thus on performance and safety.

Despite the biological mechanisms underlying culture, culture can change over time and does not carry the authority of law. Any enforcement of culture comes in the form of expectations to conform (peer pressure). The battle for culture is one of ideas. As ideas emerge and are fought over, the ones that stick and spread change the culture of an organization. The most important part of overcoming a negative cultural influence is identifying it as negative in the first place. After that, you have to make an effort to seek out information and guidance that provides you with alternatives. This may make you an outcast from your company or even department. However, you might just change the culture. Start with your own wellness commitment, and then work to get the company officer and shift mates involved.

Rules

Unlike culture, rules are expectations of action that are codified in writing and enforceable by authority. They are often manifested as policy within an organization. However, in a fire department, unofficial rules sometimes exist at the station level and are enforced by the station captain. Usually, these rules are meant to ensure the upkeep of the station or to support seniority.

Rules have a major impact on individuals and entire workforces. For example, the absence of weigh-in rules for firefighters has contributed to obesity rates in the fire service of 35%–45%, which is the same or slightly higher than the general public.[36] The military has weigh-in rules and only has a 12% obesity rate.[37] Shift schedules are another example of policy that can affect you. Your shift schedule may require you to stay at work for 24 hours, which can impact your sleep, exercise, and nutrition.

Rules are influenced by culture, and as culture changes, the rules are often updated. You do not have to follow culture, but you generally do have to follow rules (you should not take this to mean that you must follow unlawful rules or orders). Until rules change, you have to adapt to the situation. If your department does not have weigh-ins, and you believe it affects your behavior or safety, consider changing departments. If you are losing sleep on duty because you work 24-hour shifts at a busy station, make sure you prioritize your sleep while off duty. Social support can help too. It's likely that someone in your network has adapted to the same challenges and can probably give you advice on how to overcome the challenge or deal with it in a constructive manner.

Resources

Resources refer to people, money, supplies, or other assets that can be used to support a goal. For fire departments, resources will affect staffing, education and training, and equipment. Your ability to engage in certain activities might change depending on your department's resources. The most important resource, especially if you lack money, is knowledge.

Each fire department has a budget and must decide on how to spend the money they have. According to the Fifth Needs Assessment of the U.S. Fire Service, which was done in 2021, 73% of fire departments did not provide a program to maintain basic health and fitness.[38] It is possible that these programs do not exist because firefighters do not know enough about health and wellness to develop programs. The problem appears to be more significant as the size of the department gets smaller. Of the departments that claimed to have an in-house health and wellness program, less than 50% offered education on general wellness, behavioral health, and physical health, which are all features of NFPA 1500 and NFPA 1583.

Rules directly impact the way an organization acquires and allocates resources. For example, having more resources might provide for better gym equipment, but engaging in exercise doesn't require lots of resources. When it comes to exercise, there are many creative ways to work out at the fire station without a gym. The internet is full of creative circuit workouts using tools from the job. There are also programs that exist to support fire departments, such as the Federal Emergency Management Agency's (FEMA) Assistance to Firefighters Grants (AFG) program.[39] The AFG program can be used to purchase needed training programs and equipment.

Conclusion

Remember that health and wellness are not the same things. Wellness is what you do, and health is what you are. Wellness is the input; health is the outcome. The concept seems simple enough that you might expect that fire service institutions are explaining this relationship to people and educating them on how to manage the relationship between wellness and health. However, there is evidence that this is not the case. Both the fire service and general public are dying from the same Big Three chronic diseases—heart disease, cancer, and suicide. Barring mandates on weight limits or fitness tests for health insurance or fire service employment, these trends are likely to continue. The Big Three are linked to an individual's lifestyle and can be improved with proper physical wellness, nutritional wellness, and mental wellness efforts. This requires initial and continuing education, as well as proper support services, for the individual and organization.

Questions

1. What is Will Durant's quote that paraphrases Aristotle's teachings?
2. What does Durant's quote mean?
3. How is Durant's quote applicable to health?
4. What is the Merriam-Webster definition of health?
5. What is the relationship between health and wellness?
6. What is resilience?
7. What are the problems with the first two common definitions of resilience (i.e., "the ability to bounce back" and "having more in the tank")?
8. How do you become more resilient?
9. What are the Big Three?
10. What is the relationship between lifestyle and the Big Three?
11. Which NFPA standards cover firefighter health and wellness?
12. What areas are covered in chapters 11 and 12 of NFPA 1500?
13. What areas are covered in chapters 6, 7, and 8 of NFPA?

14. What is OSHA's main training mandate?
15. What is the Public Safety Officers Benefit Act (PSOB)?
16. Based on what the PSOB covers, what is the minimum that firefighters should be taught regarding health and wellness?
17. What are Aristotle's four assessments to determine full responsibility?
18. How are Aristotle's four assessments applied in the Change Assessment Model?
19. What is the firefighter wellness deficit?
20. What percentage preventability rate is a constant theme for the Big Three?

Notes

1. Rita F. Fahy, *U.S. Firefighter Fatalities Due to Sudden Cardiac Death, 1995–2004* (National Fire Protection Association, 2005); Elpidoforos S. Soteriades et al., "Cardiovascular Disease in U.S. Firefighters: A Systematic Review," *Cardiology in Review* 19, no. 4 (2011): 202–15, https://doi.org/10.1097/CRD.0b013e318215c105.
2. Justin Yang et al., "Sudden Cardiac Death Among Firefighters ≤45 Years of Age in the United States," *The American Journal of Cardiology* 112, no. 12 (2013): 1962–67, https://doi.org/10.1016/j.amjcard.2013.08.029.
3. Merriam-Webster, s.v. "health (n.)," accessed July 14, 2023, https://www.merriam-webster.com/dictionary/health.
4. Merriam-Webster, s.v. "resilience (n.)," accessed July 14, 2023, https://www.merriam-webster.com/dictionary/resilience.
5. Merriam-Webster, s.v. "resilience (n.)"; Dictionary.com, s.v. "resilience (n.)," accessed July 14, 2023, https://www.dictionary.com/browse/resilience.
6. Peter Rapp and Molly Wagster, "Cognitive Reserve Research Reaches for the STARRRS," *Inside NIA*, July 19, 2019, https://www.nia.nih.gov/research/blog/2019/06/cognitive-reserve-research-reaches-starrrs; "Home," Reserve and Resilience, accessed July 14, 2023, https://reserveandresilience.com/; Eleanor Balme, Clare Gerada, and Lisa Page, "Doctors Need to Be Supported, Not Trained in Resilience," *BMJ* 351 (2015): h4709, https://doi.org/10.1136/bmj.h4709; Carolyn Emily Schwartz et al., "Is the Link Between Socioeconomic Status and Resilience Mediated by Reserve-Building Activities: Mediation Analysis of Web-Based Cross-Sectional Data from Chronic Medical Illness Patient Panels," *BMJ Open* 9 (2019): e025602, https://doi.org/10.1136/bmjopen-2018-025602.
7. "Resilience," American Psychological Association, accessed July 14, 2023, https://www.apa.org/topics/resilience; Andrea Ovans, "What Resilience Means,

and Why It Matters," *Harvard Business Review*, January 5, 2015, https://hbr.org/2015/01/what-resilience-means-and-why-it-matters; David Fletcher and Mustafa Sarkar, "Psychological Resilience: A Review and Critique of Definitions, Concepts, and Theory," *European Psychologist* 18, no. 1 (2013): 12–23, https://doi.org/10.1027/1016-9040/a000124; Steven M. Southwick et al., "Resilience Definitions, Theory, and Challenges: Interdisciplinary Perspectives," *European Journal of Psychotraumatology* 5, no. 1 (2014): 25338, https://doi.org/10.3402/ejpt.v5.25338.

8. Bradley J. Cardinal et al., "If Exercise Is Medicine, Where Is Exercise in Medicine? Review of U.S. Medical Education Curricula for Physical Activity-Related Content," *Journal of Physical Activity & Health* 12, no. 9 (2015): 1336–43, https://doi.org/10.1123/jpah.2014-0316.
9. Gamaliel Baer, "Initial and Continuing Physical and Behavioral Health and Wellness Education in the Fire Service: An Innovation Study on Heart Attack, Suicide, and Cancer Prevention for Howard County Fire and Rescue," (PhD diss., University of Southern California, 2019).
10. "10 Leading Causes of Death, United States," WISQARS: Leading Causes of Death Visualization Tool, accessed July 14, 2023, https://wisqars.cdc.gov/data/lcd/home.
11. Baer, "Initial and Continuing Physical and Behavioral Health."
12. Soteriades et al., "Cardiovascular Disease"; Yang et al., "Sudden Cardiac Death."
13. Occupational Safety and Health Administration, *Training Requirements in OSHA Standards* (OSHA 2254-09R, U.S. Department of Labor, 2015), https://www.osha.gov/sites/default/files/publications/osha2254.pdf.
14. "The Human Genome Project," National Human Genome Research Institute, last modified September 2, 2022, https://www.genome.gov/human-genome-project.
15. Steven L. Salzberg, "Open Questions: How Many Genes Do We Have?" *BMC Biology* 16 (2018): 94, https://doi.org/10.1186/s12915-018-0564-x.
16. David L. Molfese, "Advancing Neuroscience Through Epigenetics: Molecular Mechanisms of Learning and Memory," *Developmental Neuropsychology* 36, no. 7 (2011): 810–27, https://doi.org/10.1080/87565641.2011.606395.
17. Simone Johnson, "The Top 10 Most and Least Stressful Jobs," *Business News Daily*, February 21, 2023, https://www.businessnewsdaily.com/1875-stressful-careers.html.
18. Antonios J. Tsismenakis et al., "The Obesity Epidemic and Future Emergency Responders," *Obesity* 17, no. 8 (2009): 1648–50, https://doi.org/10.1038/oby.2009.63; Marnie Dobson et al., "Exploring Occupational and Health Behavioral Causes of Firefighter Obesity: A Qualitative Study," *American Journal of Industrial Medicine* 56, no. 7 (2013): 776–90, https://doi.org/10.1002/ajim.22151; Denise Smith et al., *Heart to Heart: Strategizing an Evidence-Based Approach to Reduce Cardiac Disease and Death in the Fire Service* (National Fallen Firefighter Foundation, 2016), https://www.skidmore.edu/responder/documents/Heart-to-Heart-WP2016.pdf; Walker S. C. Poston et al., "The Prevalence of Overweight, Obesity, and Substandard Fitness in a Population-Based Firefighter

Cohort," *Journal of Occupational and Environmental Medicine* 53, no. 3 (2011): 266–73, https://doi.org/10.1097/JOM.0b013e31820af362.
19. Aristotle, *Nicomachean Ethics*, trans. W. D. Ross (2005), 1110a, http://classics.mit.edu/Aristotle/nicomachaen.html.
20. Preetha Anand et al., "Cancer Is a Preventable Disease that Requires Major Lifestyle Changes," *Pharmaceutical Research* 25, no. 9 (2008): 2097–116, https://doi.org/10.1007/s11095-008-9661-9.
21. Farhad Islami et al., "Proportion and Number of Cancer Cases and Deaths Attributable to Potentially Modifiable Risk Factors in the United States," *CA: A Cancer Journal for Clinicians* 68, no. 1 (2018): 31–54, https://doi.org/10.3322/caac.21440; GBD 2019 Cancer Risk Factors Collaborators, "The Global Burden of Cancer Attributable to Risk Factors, 2010–19: A Systematic Analysis for the Global Burden of Disease Study 2019," *The Lancet* 400, no. 10352 (2022): 563–91, https://doi.org/10.1016/S0140-6736(22)01438-6.
22. Robert D. Daniels et al., "Mortality and Cancer Incidence in a Pooled Cohort of U.S. Firefighters from San Francisco, Chicago and Philadelphia (1950–2009)," *Occupational and Environmental Medicine* 71, no. 6 (2014): 388–97, https://doi.org/10.1136/oemed-2013-101662.
23. Jonathan T. O. Cavanagh et al., "Psychological Autopsy Studies of Suicide: A Systematic Review," *Psychological Medicine* 33, no. 3 (2003): 395–405, https://doi.org/10.1017/s0033291702006943; E. Clare Harris and Brian Barraclough, "Suicide as an Outcome for Mental Disorders: A Meta-Analysis," *British Journal of Psychiatry* 170, no. 3 (1997): 205–28, https://doi.org/10.1192/bjp.170.3.205.
24. Lynette L. Craft and Frank M. Perna, "The Benefits of Exercise for the Clinically Depressed," *Primary Care Companion to the Journal of Clinical Psychiatry* 6, no. 3 (2004): 104–11, https://doi.org/10.4088/pcc.v06n0301; Mayo Clinic Staff, "Depression and Anxiety: Exercise Eases Symptoms," Mayo Clinic, last modified September 27, 2017, https://www.mayoclinic.org/diseases-conditions/depression/in-depth/depression-and-exercise/art-20046495; "Exercise Is an All-Natural Treatment to Fight Depression," *Harvard Health Publishing*, February 2, 2021, https://www.health.harvard.edu/mind-and-mood/exercise-is-an-all-natural-treatment-to-fight-depression.
25. Joshua J. Broman-Fulks et al., "Effects of Aerobic Exercise on Anxiety Sensitivity," *Behaviour Research and Therapy* 42, no. 2 (2004): 125–36, https://doi.org/10.1016/S0005-7967(03)00103-7; Elizabeth Anderson and Geetha Shivakumar, "Effects of Exercise and Physical Activity on Anxiety," *Frontiers in Psychiatry* 4 (2013): 27, https://doi.org/10.3389/fpsyt.2013.00027; Kaushadh Jayakody, Shalmini Gunadasa, and Christian Hosker, "Exercise for Anxiety Disorders: Systematic Review," *British Journal of Sports Medicine* 48, no. 3 (2014): 187–96, https://doi.org/10.1136/bjsports-2012-091287; Eric M. Hecht et al., "Cross-Sectional Examination of Ultra-Processed Food Consumption and Adverse Mental Health Symptoms," *Public Health Nutrition* 25, no. 11 (2022): 3225–34, https://doi.org/10.1017/S1368980022001586.

26. Dana Schultchen et al., "Bidirectional Relationship of Stress and Affect with Physical Activity and Healthy Eating," *British Journal of Health Psychology* 24, no. 2 (2019): 315–33, https://doi.org/10.1111/bjhp.12355.
27. Aylin Mehren et al., "Physical Exercise in Attention Deficit Hyperactivity Disorder—Evidence and Implications for the Treatment of Borderline Personality Disorder," *Borderline Personality Disorder and Emotion Dysregulation* 7 (2020): 1, https://doi.org/10.1186/s40479-019-0115-2; Yuri Rassovsky and Tali Alfassi, "Attention Improves During Physical Exercise in Individuals with ADHD," *Frontiers in Psychology* 9 (2018): 2747, https://doi.org/10.3389/fpsyg.2018.02747.
28. Mark A. Smith and Wendy J. Lynch, "Exercise as a Potential Treatment for Drug Abuse: Evidence from Preclinical Studies," *Frontiers in Psychiatry* 2 (2012): 82, https://doi.org/10.3389/fpsyt.2011.00082; Sarah E Linke and Michael Ussher, "Exercise-Based Treatments for Substance Use Disorders: Evidence, Theory, and Practicality," *The American Journal of Drug and Alcohol Abuse* 41, no. 1 (2015): 7–15, https://doi.org/10.3109/00952990.2014.976708; Dongshi Wang et al., "Impact of Physical Exercise on Substance Use Disorders: A Meta-analysis," *PLOS ONE* 9, no. 10 (2014): e110728, https://doi.org/10.1371/journal.pone.0110728.
29. Harris and Barraclough, "Suicide as an Outcome"; Yoojin Song et al., "Comparison of Suicide Risk by Mental Illness: A Retrospective Review of 14-Year Electronic Medical Records," *Journal of Korean Medical Science* 35, no. 47 (2020): e402, https://doi.org/10.3346/jkms.2020.35.e402; Silke Bachmann, "Epidemiology of Suicide and the Psychiatric Perspective," *International Journal of Environmental Research and Public Health* 15, no. 7 (2018): 1425, https://doi.org/10.3390/ijerph15071425; Judit Balazs and Agnes Kereszteny, "Attention-Deficit/Hyperactivity Disorder and Suicide: A Systematic Review," *World Journal of Psychiatry* 7, no. 1 (2017): 44, https://doi.org/10.5498/wjp.v7.i1.44.
30. Sara Veronica Fredriksson et al., "How Are Different Levels of Knowledge About Physical Activity Associated with Physical Activity Behaviour In Australian Adults?" *PLOS ONE* 13, no. 11 (2019): e0207003, https://doi.org/10.1371/journal.pone.0207003.
31. Albert Bandura, "Social Cognitive Theory for Personal and Social Change by Enabling Media," in *Entertainment-Education and Social Change: History, Research, and Practice*, ed. Arvind Singhal, Michael J. Cody, Everett M. Rogers, and Miguel Sabido (Routledge, 2004).
32. National Center on Birth Defects and Developmental Disabilities, Office of Genomics and Precision Public Health, "Family Health History: The Basics," Centers for Disease Control and Prevention, last modified May 5, 2023, https://www.cdc.gov/genomics/famhistory/famhist_basics.htm; "Why Is It Important to Know My Family Health History?" MedlinePlus, last modified May 12, 2021, https://medlineplus.gov/genetics/understanding/inheritance/familyhistory/.
33. Bridget F. Kay et al., "Assessment of Firefighters' Cardiovascular Disease-Related Knowledge and Behaviors," *Journal of the Academy of*

Nutrition and Dietetics 101, no. 7 (2001): 807, https://doi.org/10.1016/S0002-8223(01)00200-0; Dorothee M. Baur et al., "Weight-Perception in Male Career Firefighters and its Association with Cardiovascular Risk Factors," *BMC Public Health* 12 (2012): 480, https://doi.org/10.1186/1471-2458-12-480.

34. John Alexander Staley III, "The Determinants of Firefighter Physical Fitness: An Inductive Inquiry into Firefighter Culture and Coronary Risk Salience" (PhD diss., The University of North Carolina at Chapel Hill, 2008).
35. Dobson et al., "Exploring Occupational and Health Behavioral Causes."
36. Tsismenakis et al., "The Obesity Epidemic"; Poston et al., "The Prevalence of Overweight"; Michelle Lynn Wilkinson et al., "Physician Weight Recommendations for Overweight and Obese Firefighters, United States, 2011–2012," *Preventing Chronic Disease* 11 (2014): e116, https://doi.org/10.5888/pcd11.140091.
37. Tracey J. Smith et al., "Overweight and Obesity in Military Personnel: Sociodemographic Predictors," *Obesity* 20, no. 7 (2012): 1534–38, https://doi.org/10.1038/oby.2012.25; Carolyn M. Reyes-Guzman et al., "Overweight and Obesity Trends Among Active Duty Military Personnel: A 13-Year Perspective," *American Journal of Preventive Medicine* 48, no. 2 (2014), 145–53, https://doi.org/10.1016/j.amepre.2014.08.033.
38. *Fifth Needs Assessment of the United States* (National Fire Protection Association, 2005), https://www.nfpa.org/News-and-Research/Data-research-and-tools/Emergency-Responders/Needs-assessment.
39. "Assistance to Firefighters Grants Program," FEMA, accessed June 2, 2023, https://www.fema.gov/grants/preparedness/firefighters.

2
PHYSICAL WELLNESS

Over the course of this chapter, you will learn how to change your physical activity by using the Change Assessment Model from chapter 1. This chapter includes three parts. *Part 1: Nature and Nurture* describes what you cannot change (nature) and what you can change (nurture). *Part 2: Deficiency and Excess* identifies the choices you have regarding the things you can change. *Part 3: Micro and Macro* describes the relationship between smaller and larger systems that you interact with; these relationships include the cell and the body as well as the individual and the group. Your reality can be shaped by how you navigate these three areas.

Part 1: Nature and Nurture

What You Cannot Change (Nature)
You have certain biological features that you have to live with. Three of those features include the stress response system, the inevitable decline of naturally occurring human growth hormone (HGH), and your genetically inherited body type. You cannot change these three features of your biology, but it is important to understand that they exist. Acknowledging and understanding these three features will give you a better appreciation for the things that you can control about your biology.

Stress response system
Stress is a part of life. As a firefighter, it is a major part of your job.[1] You are constantly dealing with varying levels of stress whether you realize it or not. Whether the stress you experience is low intensity or high intensity, or whether it lasts for a short or long period of time, your body reacts to that stress. The reaction, known as the stress response, is a way for your body to increase your

chances for survival. Chemicals and hormones are released that signal your body to react to the stress, for better or worse.

The stress response is often described as "fight, flight, or freeze." During an intense stress event, your body is primed by certain chemicals that prepare you to defend yourself, attack someone or something, or escape from imminent danger as fast as possible. Sometimes those chemicals are so overwhelming that you can freeze momentarily. However, stress is not always a short-term life threat.[2] Stress can be a long-term event such as chronic family- or work-related issues.

Depending on the type of stress and the duration of the stress, your body will respond with signals to either grow or decay.[3] Your body actually has a default setting to decay, and there are times when decay is necessary. The default decay mechanism aids survival because when food is scarce, the ability to decay allows your body to shed unnecessary muscle and bone mass, as well as downgrade body systems to limit the use of calories. When there is plenty of food and you are engaging in life, your body can build new muscle and bone mass and spend more calories on optimizing body systems.

You cannot change the way your body reacts to stressful situations, but you can change the situations your body has to react to. As an example, exercise causes stress on your body that will result in growth as your body tries to prepare for doing that exercise again. Alternatively, sitting and doing nothing all day causes stress on your body that will result in decay as it prepares for what it thinks is hibernation. The key is knowing which activities trigger the growth response, so that you can give yourself the best chance at staying healthy and young for as long as possible.

Human growth hormone decline

As of the creation of this book, humans are still bound to a finite lifetime. As you age, a certain amount of decay is inevitable, and eventually this decay will cause you to die. If it is any consolation, there are very smart people working on ways to delay the natural aging process, and even possibly reverse it. Perhaps one day, decay-related aging will be a choice, not a rule. However, until that time, it is useful to know that your body experiences a decrease in naturally occurring HGH over time that looks something like figure 2–1.[4]

In figure 2–1, the y-axis displays HGH levels, and the x-axis displays age. The curvy line represents the maximum amount of HGH you can get at that age. But there is another issue. From the time you are born until the time you are finished physically maturing, your body is flooded with HGH. Think of that time as the "Free Ride" phase of your life. The "Free Will" phase begins after you are finished physically maturing. The Free Ride phase is a very forgiving time. You can physically engage in many rigorous activities and recover and repair very easily. You can also do very little physical activity and still grow to maturity. While you are in the Free Ride phase of your life, your daily HGH would look something like figure 2–2.

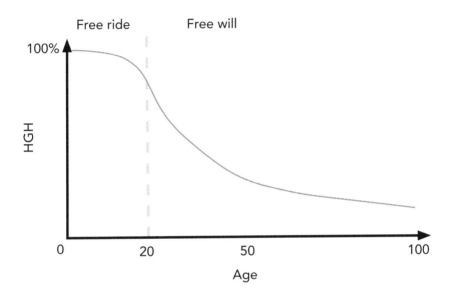

FIGURE 2–1. Maximum HGH declines with age

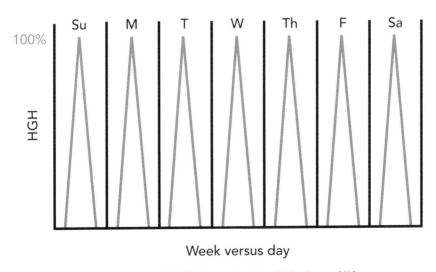

FIGURE 2–2. Weekly calendar of HGH during the Free Ride phase of life

After you are done physically maturing, naturally occurring HGH steadily decreases over time until you die. This Free Will phase is a much less forgiving time. However, there are ways to optimize your body's HGH, which will be covered in *Part 2: Deficiency and Excess*. For now, just know that as an adult you can activate HGH in your body, but only if you put yourself in situations where the stress response is activated in a way that causes HGH to be triggered. If you do not activate HGH in your daily life, you are missing an opportunity to stay as youthful and healthy as possible for as long as possible. Figure 2–3 compares the baseline levels of HGH you get during the Free Ride phase (100%) and the baseline levels of HGH you get during Free Will phase (Adult baseline).

If you accept that there is a Free Ride phase and a Free Will phase of HGH in your life, then it is worth asking yourself a question. Which days of the week would you be willing to block HGH from your children, your nieces or nephews, or yourself as a child? If your answer is "none of them" (which hopefully was your answer), and if you accept the idea of a Free Will phase of life where you can change your HGH output, why would you accept anything less than maximizing your daily HGH as an adult? The wave of life eventually crashes, but the more HGH you produce daily, the higher up on the wave you can ride through the entire aging process. The less HGH you produce daily, the more the wave is crashing on you and the faster you will disintegrate.

Genetic inheritance: body type

Your genes determine the type of body frame you inherit from your biological parents. Body type is not the only genetic inheritance you receive (e.g., cognitive ability, predisposition to diseases), but for the purposes of this chapter, it

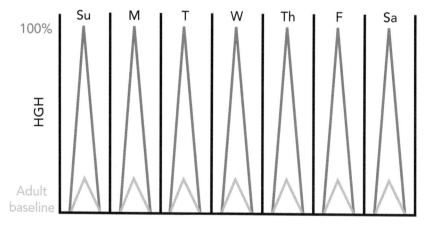

FIGURE 2–3. Weekly Calendar of Free Ride HGH (green) vs. Free Will HGH (orange)

is the focus. You should not be spending a lot of energy, time, and money trying to fundamentally change your inherited genetic code. It is important to realize that no amount of fitness training will change your body type.

Dr. William Sheldon introduced the idea of three main body types in the 1940s, and these categories are still used today for fitness education.[5] The three main body types as shown in figure 2–4 are:

1. Ectomorph: Lean and long with a thin frame and less muscle mass
2. Mesomorph: Strong but moderate frame and musculature; often considered the "Goldilocks" type
3. Endomorph: Large skeletal frame with lots of muscle mass

There is a spectrum of body types, with ectomorph, mesomorph, and endomorph types being identifiers on that spectrum. It is normal not to fit perfectly into one body type. However, knowing where you are on the body type spectrum will give you a better idea of how your body can realistically change. For example, weight gain tends to be harder the closer one is to the ectomorph body type and easier the closer one is to the endomorph body type. Body type identification can help you understand what you can and cannot change.

Firefighters are industrial athletes. Just like in sports, athletes can come in all shapes and sizes. In American football, wide receivers, running backs, and offensive linemen all have stereotypical body types. The wide receiver is longer and thinner (ectomorph). The running back is usually not as tall as a wide receiver but is more muscular (mesomorph). The offensive lineman is often tall

FIGURE 2–4. Ectomorph, mesomorph, and endomorph body types

and thick (endomorph). Each position requires athletic capabilities that favor the strengths of one body type over another. However, the athletes that play in each position are all expected to be operating at a high level of performance. Firefighters are similar and come in all shapes and sizes. You cannot change your body type, but you can still perform at your highest level.

Body type is not the same thing as body mass index (BMI) or body fat. In other words, obesity is not a body type! BMI is simply your weight in kilograms divided by your height in meters squared (kg/m^2). While BMI is often used to estimate body fat, it does not actually measure body fat. BMI is a way of calculating the height-to-weight ratio which provides a general idea about your mass. BMI does not distinguish between muscle mass and fat mass. Therefore, BMI can change if your muscle mass changes, if your body fat changes, or if both change. For example, if you gain a lot of muscle, your BMI can go up even if you lose body fat. Table 2–1 presents the body fat percentage and BMI thresholds based on the American Council on Exercise (ACE) Health Coach Manual.[6]

Body fat percentage is the amount of fat you have on your body relative to your total weight. Since body fat is used to store energy, you can reduce body fat by taking in less energy (consuming less calories) or by expending more energy (burning more calories). Depending on your body type, BMI may be harder to change than body fat percentage. An endomorph (lineman) body type with lots of muscle mass will generally be higher on the BMI scale even if they have a low body fat percentage. This presents a problem when BMI is the only metric used to determine whether someone is overweight or obese. However, for most of the population BMI is a close estimate. BMI is also easy and cheap to determine because all you need is someone's height and weight.

There is something known as the fat mass and obesity associated (FTO) gene, which is also called the "obesity gene," and everyone has this gene. If you inherit two copies of the high-risk version, you will gain weight much easier than someone who has either one copy of the high-risk version or two copies of the low-risk version. However, even if you have two copies of the

TABLE 2–1. Body fat percentage and BMI thresholds

Body fat percentage			Body mass index	
Classification	Men BF%	Women BF%	Classification	BMI
Essential	2–4	10–12	Underweight	<18.5
Athletes	6–13	14–20	Normal	18.5–24.9
Fitness	14–17	21–24	Overweight	25.0–29.9
Acceptable	18–25	25–31	Obese	30.0–40.0
Obese	>25	>31	Extremely obese	>40.0

high-risk version, doing 300 minutes of moderate exercise (like walking) each week will virtually zero out the impact of a double high-risk FTO gene completely.[7]

Firefighters are a good representation of the general public when it comes to being overweight and obese. One study on firefighter BMI and body fat found that firefighters might even be slightly more overweight and obese than the general public.[8] In the study, firefighters believed that BMI would overestimate body fat and underestimate muscle mass, thus providing misleading data. However, the researchers actually found that BMI underestimated body fat in the firefighter study group. In other words, firefighters on average were carrying more fat than what BMI metrics suggested.

BMI, and more precisely body fat, can be controlled through lifestyle adjustments. Shortly after the firefighter obesity study was published, two studies found that the military obesity rate was only 12%.[9] Unless you believe that military personnel have thinner builds on average (i.e., ectomorph) compared to firefighters (i.e., mesomorph or endomorph), then it is likely that the lower obesity rate in the military is due to the mandatory weight limits that the military imposes on its members. If you assume that military personnel and fire service personnel are likely similar in body types, then the good news is that the fire service could also achieve a 12% obesity rate. The bad news is that without mandatory weight limits, achieving 12% for the entire fire service will be difficult.

Even if your fire department never imposes mandatory weight limits for employment, there is a good reason that you should keep your body fat down. Figure 2–5 shows a chart of the correlation between body fat and HGH. As

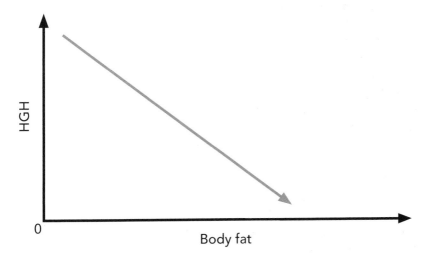

FIGURE 2–5. Body fat lowers HGH

you accumulate more body fat, you will release less HGH on a daily basis.[10] This relationship appears to hold true whether you are young or old and whether you are male or female. Carrying excess body weight also reduces your body's ability to heal and recover because it increases generalized inflammation.[11] Therefore, regardless of your body type, carrying excess body fat will negatively impact your HGH production and overall health.

What You Can Change (Nurture)

As a firefighter, you owe it to yourself, your family, your shift, and those you serve to be as healthy and fit as possible for as long as possible. Now that you are familiar with a few factors that you cannot change (stress response, decline in HGH, and body type), you need to understand the physical activities that you can leverage to optimize your HGH production and overall health. The spectrum of physical activity includes more than just exercise, as shown in figure 2-6.

> NFPA 1500 and NFPA 1583 provide standards for physical fitness, general wellness, and health promotion education. Those standards can be found in NFPA 1500 Chapters 4 (Section 4.3), 11 (Section 11.2, 11.3, and 11.7), and 12 (Section 12.2) and NFPA 1583 Chapters 5, 6, 7 and 8.

Your body requires energy to stay alive. So, unless you are cryogenically frozen, you are either consuming calories or you are operating off of stored calories. You are either unconscious (e.g., sleeping) or conscious (e.g., firefighting). While unconscious, you can be in a coma or asleep. Thanks to technology, we can track what stage of sleep you are in by observing brain waves.[12] While conscious, you are either sedentary (e.g., sitting) or active (e.g., walking). If you are reading a book, sitting and talking, or even meditating, you are in a sedentary state. If you are active, your activity is classified as light, moderate, or vigorous. The next section will provide an overview of exercise, fasting, and sleep, all of which can optimize your HGH production naturally.

Exercise

The first area of physical activity you can control is exercise. Exercise is the act of exerting your body to develop or maintain physical fitness. In figure 2–6, exercise exists under the "Active" portion of the flowchart and ranges from light to vigorous. As you will learn, even vigorous activity has a wide spectrum.

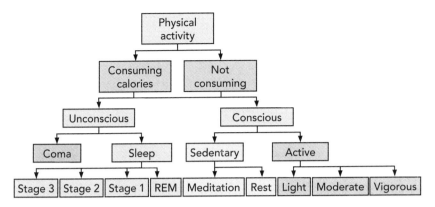

FIGURE 2–6. Physical activity spectrum

Firefighters are considered industrial athletes because of the amount of potential exertion required from the job, which can include very vigorous activity.

Firefighters should care about exercise. Being in good shape helps you to do your job better. But from a wellness standpoint, you should know that at a certain intensity level you will release HGH during and after exercise.[13] You can exercise all you want below a certain threshold, and you will never release significant HGH that way.

Fasting

The second area of physical activity that you can control is when you choose to refuel. Fasting is the act of abstaining from food or drink. Fasting can occur while you are awake and even while you are active, and it also occurs every time you sleep. When you are sleeping, you are fasting, which is why the first meal after waking up is called breakfast ("breaking the fast"). Even though you do not eat while you sleep, you are not making a conscious effort to fast, and because you are unconscious, you do not notice any hunger nor are you bothered by it.

Choosing to fast while you are conscious is a different idea entirely. The reason is because it may involve some discomfort in the form of hunger. However, some discomfort also occurs when you exercise appropriately. If you choose to exercise, you accept the discomfort knowing that there is a positive effect on your body. You can also receive a positive effect from fasting.

In order to understand why fasting is important, you need to understand how your body stores and uses fuel. Your body can only store a limited amount of energy in the form of sugar in your muscles and liver.[14] Your liver, being a fraction of the size of your body's muscle mass, can only store a fraction of the sugar that your muscles can. However, you can store an unlimited amount of energy in your body fat.

In the book *Younger Next Year*, Dr. Henry Lodge explains that depending on the physical exertion you are engaged in, your body will use different fuels.[15] The reason is that some fuels are biologically more expensive than others and it does not make sense to use what is not needed. The three main fuels your body uses are fat (triglycerides), sugar (glucose), and creatine phosphate.

During low exertion periods, your body is designed to primarily rely on fat for fuel. Fat is used in the form of triglycerides. What is not used gets stored in the liver or long-term fat to be used later. However, only a few triglyceride molecules can make it through a capillary at a time. This limitation creates a mechanical bottleneck. When you enter an exertion level that mimics hunting or being hunted and your energy demands increase, your body can no longer rely only on triglycerides because they cannot provide fuel fast enough. Instead, your body begins to spend the sugar (which is stored in your muscles as glycogen) in addition to burning fat (triglycerides). If you engage in a live-or-die scenario in which you need to exert yourself at 100%, then for about 10 seconds or so your body spends creatine phosphate in addition to sugar and fat.

Your body uses and stores energy to give it the best chance of survival. There is no point in using sugar that is stored in your muscles if you do not need to use it. The same goes for creatine phosphate. However, in order for you to respond to a life threat and engage in fight or flight, sugar and creatine phosphate must always be ready to be used. Your body is always working to top off those fuels as soon as possible when they are not in use. Your muscles fill up with sugar automatically. Your liver does the same. Creatine phosphate also is stored automatically and can be replenished in as little as 3–5 minutes.[16] However, after a certain point, your body cannot store any more sugar in its short-term storage areas (muscles and liver).

In order to safely store excess sugar when your muscles and liver are topped off, your body will store that sugar as fat in your body fat. In other words, if you have not engaged in any type of strenuous activity for a significant amount of time, then your muscles and liver are likely topped off with sugar and any food you eat will go directly to fat storage. Figure 2–7 presents a simplified graph of the three main storage areas of energy from sugar.

Similar to a fuel tank in your vehicle, once your muscles are filled to capacity with sugar, that is it. The liver is like a reserve fuel tank that you might see on off-road vehicles. If your muscles and liver are filled with sugar stores, then any excess energy you consume that gets converted to sugar will be stored in your body fat! Having lots of excess body fat would be like towing a trailer of

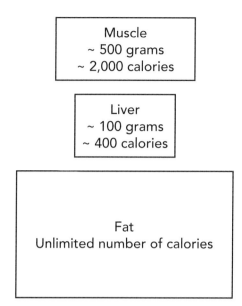

FIGURE 2–7. Sugar storage systems in our body

extra fuel behind your vehicle, even though the main fuel tank is full and you have a full reserve tank.

Just as a fuel tank can stay filled with fuel until it is used, the same is true with sugar in your muscles. Short-term fasting as an activity will not cause you to use sugar from your muscles. The only activity that will cause sugar to decrease from your muscles is exercise beyond a certain threshold of intensity. However, your liver lets out a steady supply of sugar in your blood so that your brain has energy to work. This is because your brain relies mostly on sugar for functionality.[17] For the liver to continue to replenish itself, you either need to eat or convert stored fat into sugar. The process of converting stored fat into sugar is called gluconeogenesis (making new sugar). Fasting will eventually prompt the body to secrete HGH, which signals the body to convert stored fat into sugar.[18] If you exercise while fasting, and you do not eat after the exercise, your muscles will eventually replenish their sugar stores through the same process. Like exercise, the HGH from fasting can optimize your health.

Sleep

The third area of physical activity that you can control is sleep. While you cannot expect to control anything regarding nighttime sleep during a shift, you can incorporate "safety naps" during the workday if your officer in charge allows it. However, at a busy station, even naps may not be possible. That is why it is essential for you to control your sleep on your off-duty days.

Sleep is an activity that firefighters should be interested in and understand. That is because while you are asleep, HGH can be secreted.[19] However, the key to HGH secretion during sleep is getting into the right stage of sleep, which does not always occur at work or during naps. In fact, depending on your sleep hygiene, you may not be getting into the right sleep stage for enough time even while you are off duty and at home in your own bed. With a little knowledge and some effort, sleep can give you lots of HGH and help you to optimize your health.

Part 2: Deficiency and Excess

Exercise

Here is the secret about exercise: it does not matter which exercise you do as long as you are exercising enough days per week (frequency [F]), for enough time (duration [D]), and with enough effort (intensity [I]). Each person is different, so "enough" will depend on your needs. Those three factors represent exercise volume ($V = F \times D \times I$) as shown in Table 2–2.[20] The goal of this section is to provide a guide on navigating the spectrum of deficiency and excess for the factors of exercise volume.

As long as you are alive, you are using energy. In exercise speak, the term *metabolic equivalent of task* (MET) is used to describe energy use. Although the MET system is universal, if necessary, it can be precisely applied to an individual. The measurement of 1 MET is your exertion level while at rest (e.g., sitting). Doing an activity that is 10 METs means you are exerting 10 times the amount of energy compared to what you exert while you are at rest.

Genetics, physical conditioning, and age will all contribute to individual MET differences. For example, if there are two firefighters of the same age and height, a firefighter who is obese will likely be exerting more energy to jog than a firefighter who is not obese. Individual differences can be addressed by a certified personal trainer. However, the main point is that as exertion

TABLE 2–2. Factors of exercise volume

Factors of exercise volume
Exercise frequency
Exercise duration
Exercise intensity

increases, METs increase. This relationship is the basis for understanding exercise volume.

For practical matters like exercise and sleep, the right amount is relative to your needs. Figure 2–6 presented a theoretical range of activity from the deepest sleep stage to the highest level of exercise intensity. Some people will spend more time in the sleep and sedentary areas and some people will spend more time in the active areas. Interestingly, as physical activity increases, sleep quality increases as well, which suggests a pendulum-like effect.[21] Finding a virtuous middle does not always mean avoiding deficiency and excess (although sometimes it does). When it comes to physical activity and its relationship with sleep, finding a virtuous middle means that you are balanced in your approach to the spectrum of activity, which should include vigorous exercise and deep sleep. When we are out of balance, we increase the likelihood of developing stress-related health issues.

Exercise frequency

Exercise frequency is the number of times an activity occurs in a given time frame. Exercising once in an hour would mean a frequency of one, and exercising twice in an hour would mean a frequency of two. A frequency of one round of exercise each hour of the day is excessive. However, if only one round of exercise is done for the entire week, that is deficient. One round of exercise per day for a whole week would be a frequency of seven rounds of exercise for a week, which is a reasonable expectation. For this section we will focus on minutes, hours, days, and weeks as shown in Table 2–3.

You can track exercise frequency on a daily basis, but tracking exercise on a weekly basis is probably the most common. Using a shift-cycle calendar might be more helpful for firefighters. An example of a shift-cycle tracker is shared later in the chapter.

The moment we leave a state of rest, we are in an exercise or active zone. So, if a firefighter sits all day in between calls, but runs five calls in a day, we could say that the firefighter had a frequency of five rounds of exercise for that day. However, knowing how many times someone engaged in exercise is not enough information to know if it was enough exercise. To do that, we need to know the time spent exercising. We need to track exercise duration.

TABLE 2–3. Minutes, hours, days, and weeks

Hour	Day	Week
60 minutes	24 hours	7 days

> Firefighter Lazy-Boy chooses to stay sedentary all day (24 hours) every week. Lazy-Boy's frequency of exercise would be 0 for every hour, zero for every day, and zero for the week because he never enters even the light exercise zone. Firefighter Iron-Man engages in exercise six days per week and stays sedentary for one day. His exercise frequency is six out of seven days for that week.

Exercise duration
Exercise duration is the amount of time you spend doing a round of exercise. For example, if you spend 60 minutes doing one round of exercise in a day, your exercise duration is 60 minutes in 24 hours. If you do 60 minutes of exercise per day for the whole week, then your exercise duration for the week is 420 minutes. Any time spent in exercise, or activity above rest, is counted as exercise duration. Firefighter Lazy-Boy, who sits all day, has a duration of 0 minutes of exercise for the day and 0 minutes of exercise for the week.

Each day has 1,440 minutes (60 minutes × 24 hours). A MET score of 1,440 is the lowest score you can have while you are alive (sleep is counted as 0.9 METs, but for the purpose of making the math simpler, we will use 1,440 as the lowest score). One MET is the unit of energy expenditure at rest, and so a firefighter who is sedentary all day has a MET score of 1,440.

> Since a 1,440 MET score is the lowest you can possibly have, that score can be used to identify exercise deficiency because it means that no exercise was done that day. Note that not resting at all for 24 hours would be excessive.

Consider a potential day at the firehouse. Say you run five calls during a 24-hour shift. If each call was 60 minutes, then you had 5 hours (300 minutes) of exercise at some level beyond rest. It is true that while running calls you will sit in a vehicle or even sit at the scene. To accurately know how much time you spent in an active state, you need to wear an activity tracker (e.g., Fitbit) or a smart watch (e.g., Apple Watch).

Any time you leave a sedentary state, you can calculate how much time (duration) was spent doing exercise. However, not all activities are created equal. Calculating the duration of time you spent on a call does not take into account whether you were standing still, doing CPR, dragging hose, or doing search and rescue. Different activities will result in different MET scores based on the level of activity. This is known as exercise intensity.

Exercise intensity

Exercise intensity refers to the physical demand an exercise puts on your body. In physics, Energy = Power × Time. In exercise, if you know the intensity and the duration, then you can know how much energy (calories) you will burn. MET scores are used to determine exercise intensity, and charts that compare different exercises can be found online.

Exercise intensity is broken into two inactive categories and three active categories. The first two categories are sleep (unconscious activity) and sedentary behavior (conscious activity), and the next three categories are light, moderate, and vigorous activity (fig. 2–8). Sleep is classified as 0.9 METs and sedentary behavior, like sitting, is classified as 1 MET.

> Light activity is from 1.5 to 3 METs, and examples include standing, slowly walking, and doing dishes. Moderate activity is 3 to 6 METs, and examples include brisk walking, pushing a lawn mower, and hiking. Vigorous activity is anything above 6 METs and includes jogging (roughly 6 METs), body weight exercise (roughly 8 METs), and jumping rope (roughly 10 METs).

Rest and exercise are both important and a wide spectrum of exercise intensity exists. As a firefighter who is expected to be able to respond to emergencies that demand high levels of activity, it would be deficient to only engage in exercises classified as light activity. On the other hand, we need rest and it would be excessive to constantly be engaged in vigorous activity.

Firefighting and gear-based activity

You should understand exercise intensity because it holds the key to a critical aspect of wellness that you can control. Recall that as you age you naturally

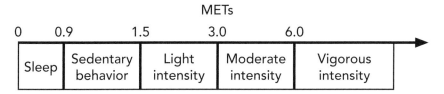

FIGURE 2–8. Exercise intensity spectrum

produce less HGH and you cannot control that. For example, at age 50 you might only be able to produce half the HGH that you could at age 25. However, your body will produce natural HGH when you exceed a certain level of exercise intensity. In other words, even though your body's maximum HGH will decline with age, you can optimize HGH at every age by engaging in exercise above a certain intensity level.

According to *NFPA 1582: Standard on Comprehensive Occupational Medical Program for Fire Departments*, firefighters need to be able to hit 12 METs on a Bruce protocol stress test for their annual physical medical evaluation.[22] At 12 METs on a Bruce protocol stress test, the firefighter is normally jogging or running uphill on a treadmill. According to a METs chart provided by the International Association of Fire Chiefs (IAFC), carrying tools upstairs in gear and self-contained breathing apparatus (SCBA) is about 12.5 METs, and search and rescue in smoke conditions is about 16 METs.[23] Activity in the 12–16 MET range is well above the minimum METs considered for vigorous activity (6 METs). Therefore, it is important to keep in mind that while walking for an hour a day every day of the week might be healthy, that level of activity does not come close to what is required as a firefighter. Even jogging at 6 METs is far lower than the METs required for fireground activities.

Activity duration and activity intensity have an inverse relationship. As exercise intensity increases (METs go higher), the time you can spend in that activity (duration) decreases. As exercise intensity decreases (METs go lower), you can spend more time doing the activity. Figure 2–9 shows the relationship between duration and intensity using 1 to 100 as a spectrum. As you get closer

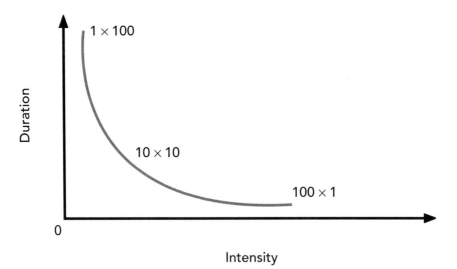

FIGURE 2–9. The trade-off between duration and intensity in exercise

to 100% intensity, the activity duration gets closer to 0 (we never reach 0% intensity until we die). An example of 100% intensity would be an all-out sprint.

The MET model of exercise intensity classifies activity as light, moderate, or vigorous. But, the classification for vigorous activity begins at 6 METs, which for most healthy people is equivalent to a jog. There is no higher classification than vigorous activity. Therefore, you can say you exercise vigorously every day of your life and never go beyond a jog. This is problematic because, under the MET model, vigorous activity includes a very wide range of intensity. So, saying you exercise vigorously can mean anything from jogging to Olympic-level training.

Most firefighters have no desire to commit MET scores to memory. And, for the purposes of knowing whether you did "enough," there is an easier method than calculating MET scores. Gear-based activity (GBA), as shown in figure 2–10, is a way of classifying your intensity level based on the metabolic gear you are in, which then determines the fuels your body uses.

There are three metabolic gears in the GBA model. Gear 1 is the lowest intensity level and can be considered the "gathering" gear. Gear 2 is a high-intensity level which begins at a jog (where "vigorous" activity begins in the MET model) and can be considered the "hunting" gear. Gear 3 is the highest intensity (all-out exertion) and can be considered the "live-or-die" gear. HGH is secreted only after you enter Gear 2.[24] It takes roughly 10 minutes in Gear 2 to begin producing HGH.[25]

> It appears that exercise can cause HGH to go as high as 300%–500% above baseline levels from getting into Gear 2. Going into Gear 3 yields even higher results.

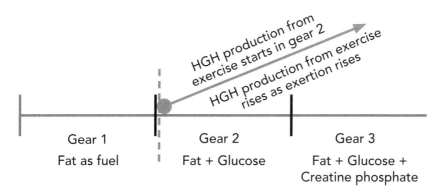

FIGURE 2–10. Gear-based activity spectrum

In Gear 1, your body uses mainly fat in the form of triglycerides as fuel. Triglycerides (fat) are constantly in the blood stream and are an efficient way to fuel your low-intensity activities. However, the width of a capillary is only about 4–8 micrometers.[26] This means that only a few triglyceride molecules (1 micrometer) can pass through a capillary at a time, which causes a mechanical limitation for energy.[27] This is similar to water flow and hose diameter. With a 1¾" handline, you can increase the flow up until about 325 gallons per minute, at which point you are maxed out. Similarly, fat as a fuel can increase up until a certain point, but eventually will be at 100% flow and can no longer increase even if your exercise intensity increases. Fat as a fuel reaches 100% flow at roughly a brisk walk, right before the intensity reaches what is known as "vigorous" exercise, which is where Gear 2 begins.

Gear 2 is the metabolic gear in which the body is using both fat and sugar as fuel. Gear 2 begins after the exercise intensity passes what is essentially a brisk walk and reaches the equivalent of a jog (roughly 6 METs, which is the threshold for vigorous exercise). The body does not stop using fat as a fuel; it simply releases stored sugar since fat has reached its fueling limit. Recall that the main storage area in the body for sugar (in the form of glycogen) is in the muscle tissue. The combination of fat as fuel and sugar as fuel continues as exercise intensity rises. Eventually, sugar also reaches a limit, which occurs when you enter into Gear 3.

Gear 3 is all-out physical exertion, which lasts for roughly 10 seconds at a time. In order to carry out the highest level of intensity, Gear 3, your body adds a third and final fuel called creatine phosphate. In Gear 3, your body uses fat, sugar, and creatine phosphate to give you the best chance of survival in live-or-die moments. However, after about 10 seconds, your creatine phosphate stores are depleted, and it takes 3 to 5 minutes for them to be restored. You can continue exercising in Gear 2, but will have to wait for your body to restore its creatine phosphate before you can reenter Gear 3.

GBA makes it easy to distinguish between different fuel systems in the body. This is important because of the secretion of HGH and its connection to the gear that your body is in. While it is healthy for people to walk and "gather" in Gear 1, the signal to produce HGH is not activated by Gear 1 activity. This matters because your body is preprogrammed to decay.

Figure 2–11 provides an analogy of the body's decay mechanism by comparing it to a Siamese clapper valve. In supply operations, we can use a Siamese clapper valve to allow for two different supply lines. Under this type of supply operation, if water is coming in from one supply line (line 2), but not the other (line 1), the clapper valve closes on line 1 so that no water is lost. As soon as water is introduced from line 1 at a high enough pressure, that water can be added into the supply system. If line 1 produces enough pressure, it can even disrupt the flow of line 2 completely.

> Without the signal to grow stronger (HGH), your body will attempt to lower the cost of maintenance by reducing muscle and bone density, which reduces caloric demand. This process allows humans to adapt to low food intake for long periods of time like during a famine.

Your body's growth and decay signals work very similarly to the example of supply lines. Line 2 in this analogy represents the body's signal to decay. In adulthood, there is always a small but steady flow of decay signals, which is the default. In order to overcome the decay signal, your body needs to activate the growth signal (HGH). Line 1 represents the growth signal in this analogy. As explained in the book *Younger Next Year*, with a strong enough growth signal, you can temporarily shut down the decay signal and force the body to repair and rebuild. However, this signal from exercise only comes if you enter Gear 2 or higher. Your body does not care which exercise you choose, as long as the intensity is high enough to enter Gear 2, which will then prompt the body to release HGH.

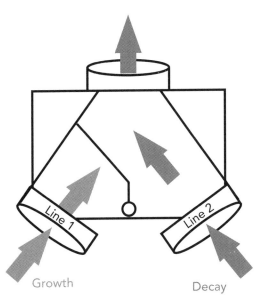

FIGURE 2–11. Your body is preprogrammed to decay in adulthood.

Therefore, the key for optimizing your health through exercise involves managing the time you spend in different gears, not the choice of specific exercises within those gears. Table 2–4 provides examples of activities for each Gear that will help show the spectrum of intensity from the bottom of Gear 1 to Gear 3. GBA is about the intensity level of what you are doing, not the specific exercise. Any exercise can be done at a low or high intensity.

> To learn more about specific exercises that firefighters can do to increase their functional fitness, consider reading *Firefighter Functional Fitness* by Dan Kerrigan and Jim Moss.

Table 2–4 aligns METs with GBA. In order for the body to secrete HGH from exercise-related activity, you have to get into Gear 2, which begins at a MET score of 6. Recall that "vigorous" activity begins at 6 METs. However, a jog does not seem to be "vigorous" compared to firefighting. A jog might be similar in exertion to an exterior fire attack while holding a handline and not moving very much. It requires energy, but it is not that hard. As you move to an interior attack, begin climbing stairs with equipment, and perhaps even engage in search and rescue, the exertion levels increase. If you have to engage in a self-rescue, in which seconds matter, you need all-out, maximum-level exertion.

> How often do you as a firefighter participate in activities at those different levels of exertion? How often do you leave Gear 1 when you are at the fire station or at home?

TABLE 2–4. Examples of Gear-Based Activity

	GBA gear 1			GBA gear 2			GBA gear 3
METs	>0 ------------------- <6			>6	12	21	<10 seconds of max effort
Natural example	Gathering			Hunting			Live or die
Exercise example	Sleeping/ meditating	Sitting	Walking	Jog	Mid-speed run	Fast run	All-out sprint
Firefighting example	Sleeping	Coffee table talk	House duties	Fighting fire	Search and rescue		Self-rescue

Aristotle explained that the mean is in between two extremes, which can be absolute or relative. For the firefighter who does not get out of Gear 1, their relative "mean" of exercise will be lower than a firefighter who gets into Gear 2 regularly. The mean of a firefighter who never leaves Gear 1 would be even lower compared to a firefighter who gets into Gear 3 regularly. However, a person cannot stay in Gear 3 all day, and unless you are running a 24-hour ultramarathon, you can't stay in Gear 2 all day.

Regarding exercise, a firefighter's mean would have to take into account the rest time and the exercise time. The more you limit your involvement in Gear 2 and Gear 3, the closer your mean will be to zero, which effectively lowers your conditioning. Without enough conditioning in Gear 2 and Gear 3, the risk of health issues increases for when you do enter those gears, especially on the fireground. Therefore, in order to have a balance regarding exercise while also benefiting from HGH, you as a firefighter need to be able to explore the whole spectrum of exercise intensity on a regular basis.

Fasting

Fasting has existed for centuries in religious practices but is now becoming a part of secular life. Fasting probably does not seem appealing in the same way that sprinting does not seem appealing. Both sprinting and fasting involve extremes and can only be sustained for a limited amount of time before your body gives out. Furthermore, for firefighters in a busy station, fasting may be undesirable or outright impractical. However, the desirability or practicality of fasting does not change the physiology of it.

After fasting for roughly 12 hours, a major metabolic change occurs as the body enters a negative energy balance.[28] The body responds to a negative energy balance by sending chemical and hormonal signals to mobilize stored energy and optimize the body for survival. Contrary to what you might expect, during short-term bouts of fasting (24 hours or less), your body will actually be primed to take on a challenge. This priming can be experienced by some people in the form of agitation and extra energy that is sometimes expressed as being "hangry."

> That "hangry" feeling would have helped your ancestors catch their next meal. But long-term fasting (which for this book will be classified as more than 24 hours) can cause your body to cannibalize your muscles for energy, and thus is not advised.

Calorie restriction

Calorie restriction is not fasting. Calorie restriction is eating fewer calories than what your body is used to consuming while avoiding malnourishment. When done intentionally, calorie restriction can be used for weight loss. However, because calorie restriction is focused on changing caloric intake, and not on the amount of time in between eating, it affects the body differently than fasting. The moment your body senses any caloric intake, HGH secretion is suppressed from the body's fasting mechanism.[29] So, even if you reduce your calorie intake to 25% of normal but you eat that food throughout the day, HGH secretion will not occur. Intermittent fasting is different from calorie restriction because, during fasting periods, there is no calorie intake at all.

Intermittent fasting

Intermittent fasting does not require you to reduce total calorie intake. During your intermittent fasting time period, you cannot consume any calories. However, it is perfectly fine to consume extra calories when you do eat in order to replace what you would have consumed otherwise.

> During fasting, it is considered acceptable to consume plain black coffee, unsweetened tea, and water, as those drinks will not register as caloric intake.

Intermittent fasting falls into two main categories. The first category is *time-restricted feeding*, which allows you to refuel each day during specific windows of time. The second category is *full-day fasting*, which involves a 24-hour fast. To avoid weight loss during full-day fasting, you would have to eat more food on non-fasting days. The following section provides examples of different types of time-restricted feeding and full-day fasting.

Time-restricted feeding

Time-restricted feeding (TRF) involves restricting the time you can eat during the day. For the purposes of intermittent fasting (IF), a 12-hour window of not eating must be included at a minimum, because that is roughly how long it takes for your body to go into a negative energy balance. As the amount of fasting time increases, the amount of time available to consume calories decreases. There are three common models of TRF, starting at 12 hours and becoming more restrictive. Those models are 12 Hour+, 16:8, and Warrior.

The 12 Hour+ model of TRF is fasting for 12 or more hours during the waking period. Under this type of TRF, you would have an early breakfast,

fast for 12 hours or more, and then have a dinner. Some religions practice daylight fasting, and depending on which part of the world you live in, the fasting can be anywhere from 10–21 hours.[30] However, most people live in areas that would require between 12–16 hours of daylight fasting. One difference is that some religious-based fasting practices prohibit liquids, including water, which is not advised if you are on duty.

The 16:8 model of TRF involves a longer fasting window. It calls for a 16-hour fast and allows an 8-hour eating window. A common approach is to fast from 8pm until noon the following day, and then refuel from noon to 8pm. A large chunk of the fasting period for the 16:8 occurs while you sleep, so the hunger is not as bothersome. During the eating period, you can have as many meals or snacks as you want, although it is common for people to stick with meal times. You can reduce the eating window to less than 8 hours in order to increase the fasting time. For example, you can do a 17:7, an 18:6, or a 19:5.

The Warrior model of TRF involves a very restrictive eating window of four hours or less, and fasting for 20 or more hours. The Warrior fast typically only allows for one large meal. This type of fast can maximize the HGH secreted during the fasting period while still allowing a refueling time each day. However, the Warrior model of TRF makes consuming the number of needed calories for the day harder, and you may feel like you are stuffing yourself for a few hours straight.

Full-day fasting
Full-day fasting involves 24 hours (or more) of not eating. Similar to TRF, the goal is not to reduce caloric intake. So, for each 24-hour fast-day in a week, you will need to increase the calories you consume on eating days in order to maintain your weight. There are two main types of full-day intermittent fasting, which are periodic fasting and alternate day fasting (ADF).

Periodic fasting involves full-day fasting for one or two days per week, then eating more on days that you do eat. A popular model of periodic fasting is called the 5:2. The 5:2 calls for two 24-hour fasts each week, and then eating as much as you want for the other five days. For periodic fasting, the one or two days can either be the same each week or can change depending on your schedule. Keep in mind that if you choose to fast one day per week, the amount of extra food you can consume on your eating days will be different than if you fast for two days a week.

ADF is a regimented version of periodic fasting. With ADF, you eat as much as you want one day and fast the next. Fasting with this method in the strictest sense means the days that you are fasting change every other week. However, you could choose to fast Monday, Wednesday, and Friday, and have the weekend off each week, as a modified version.

Firefighting and fasting

Firefighters can certainly incorporate fasting into their lives and still be functional and energized at work. Firefighters will find it easy to incorporate TRF into their workday which guarantees time to eat. For firefighters who are on 24-hour shift-work schedules, periodic fasting (5:2) or a modified schedule of ADF could be implemented to align with off days, but should not be tried during the workday.

> As a cautionary note: if you have been diagnosed with any medical conditions requiring medication, you should talk to your doctor or dietician before fasting.

While fasting on a shift, muscle glycogen storage levels are important to keep in mind. Recall that muscle glycogen stores will refill even in a fasting state. However, they refill fastest when consuming food. Therefore, if you are fasting on shift and respond to a box-alarm that drains your energy, you may want to consider breaking your fast to refuel. This is especially true if you work at a station that has a good chance of running a high number of calls.

Refueling after a workout can be planned even if you are doing TRF. For example, by using the 16:8 TRF method, you can come to work in a fasting state and exercise before noon. Assuming your refueling time is from noon to 8:00 p.m., you can eat anytime in that window. Alternatively, if you are doing 12 Hour+ (daylight) fasting or Warrior fasting, you can make a decision on whether you need to refuel after an emergency response. If you are fasting but have not exercised or responded to a physically taxing call, your muscles are filled with glycogen. So, you are good for at least one physically demanding call. After that call, you will have to decide whether to break your fast and refuel immediately or wait until your next refueling period.

Fasting exists on a spectrum but is not a measurement of intensity like exercise. It is not a question of having fewer calories. Fasting is the absence of eating, and so there is only one intensity level; you cannot eat at all. But, the feeling of hunger becomes more intense the longer you fast. As mentioned earlier, a period of 12 hours appears to be the minimum fasting time for HGH secretion to begin. The further you go beyond the threshold, the more HGH you secrete, as shown in figure 2–12.

> The 12-hour threshold for HGH is similar to the threshold between Gear 1 and Gear 2 for exercise. If you do not cross the threshold, you will not get HGH from fasting.

Fasting time and HGH secretion have a positive relationship. In other words, as you fast longer, your body will secrete more HGH. It appears that fasting can cause HGH to increase anywhere from 300% to 2000% above baseline in just 24 hours.[31]

Another study found that HGH secretion can reach its maximum impact after skipping just the first two meals of the day, which suggests that a Warrior fast is just as effective as a 24-hour fast.[32] Additionally, it appears that fasting during the day releases more HGH than fasting during the night, which is probably because during the day you are burning calories faster than when you are sleeping, which results in larger calorie deficits.

When intermittent fasting, the idea is not necessarily to reduce total caloric intake. Even if you fast for 24 hours, your total caloric intake for the week can remain the same, or even increase, if you have more calories during days when you consume food. Instead, the idea of fasting is to provide time for your body to digest and also to allow your body to focus on other functions like repairing tissue. In addition to secreting HGH, other benefits can come from fasting. After about 18 hours of fasting, your body begins a process called autophagy, which means self-eating.[33]

> If fasting had "gears" like GBA, autophagy would occur in Gear 3 of fasting. During autophagy, the body "eats" any old, dying, or even cancerous cells and uses them for energy.[34]

Firefighters can benefit from fasting as a way to naturally increase HGH. Like exercise, fasting can be practiced in a safe way that exists in the mean between deficiency and excess. Based on the available research, fasting for less than 12 hours is deficient if your goal is HGH secretion. However, for the purposes of increasing HGH, fasting for more than 24 hours at a time is excessive

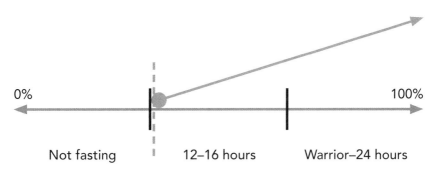

FIGURE 2–12. Intermittent fasting spectrum

and not recommended. There are multiple models to choose from that firefighters can use on or off duty.

Sleeping

Sleep is an important part of health. It is a time when your body can repair and grow. Lack of sleep not only results in cognitive impairment, but also increases stress.[35] However, more sleep is not necessarily better. A meta-analysis that included 35 studies and over 1.5 million participants found the optimal sleep time for health to be 7 hours.[36] Other studies use 6–8 hours as the target.[37]

> The studies show that the farther you get from the 7-hour target, whether it is less sleep or more sleep, the higher your chances of all-cause mortality.

While you sleep, you cycle through different stages as shown in table 2–5.[38] A typical sleep cycle is roughly 90 minutes, but can change throughout the evening. Additionally, as the night goes on, the time spent in the deep sleep stage (stage 3) shortens and the time spent in rapid eye movement (REM) sleep gets longer. Deep sleep (stage 3) and REM sleep are critical to your physical and mental health because of the benefits you receive in those stages. In stage 3, your body secretes HGH.[39]

> It appears that deep sleep can cause an increase in HGH of 850% from baseline on average, and even up to 1800%.

In REM sleep, your body organizes and manages memories.[40] To stay healthy, firefighters should be trying to maximize the time we spend in stage 3 and REM sleep. If your sleep is interrupted just before or during stage 3 or REM sleep (or both), you risk interfering with major health benefits. As a

TABLE 2–5. Sleep stages

Sleep stages	Time (minutes)	Benefit
Stage 1 (light)	1–5	Rest
Stage 2 (light)	10–25	Rest
Stage 3 (deep)	20–40	HGH
Rapid eye movement (REM)	10+	Memory storage

firefighter, disrupted sleep is part of your job while on duty. To make matters more serious, as you age, you spend less time in deep sleep (stage 3), which is correlated with lower production of HGH. Therefore, as a firefighter, you should be more interested than the average person on how sleep works and how to optimize the sleep that you can control while you are off duty.

Recall that exercise volume ($V = F \times D \times I$) is a product of frequency, duration, and intensity. Sleep is similar. Frequency is how often you get sleep, duration is how long you sleep for, and intensity is how deeply you sleep. Your health relies on having enough sleep volume, just as it relies on having enough exercise volume. In fact, sleep is enhanced when you exercise![41] However, exercise is not the only action you can take to improve your sleep.

Sleep hygiene refers to the habits that you develop regarding your sleep. There is a lot of free information on the internet about ways to improve your sleep by changing things within your control. Some of the free information is marketing to get you to spend money on a costly new bed or monthly subscriptions to supplements. However, some things you can control do not even require you to sleep in a bed and are cheap or free.

Sleep improvement methods

Table 2–6 provides a table of research-supported ways in which you can improve your sleep quality and maximize your stage 3 sleep. The table focuses on controlling certain daytime activities, your sleep environment, and your sleep schedule. Specific details of ways to improve sleep are not the focus of this book, but can be found in the research cited for the examples given in the table.

> Stage 3 (deep) sleep is when you secrete HGH, and there are evidence-based methods to optimize sleep in general, and stage 3 sleep in particular.

TABLE 2–6. Research-supported sleep improvement methods

Research-supported sleep improvement methods			
Daytime activity	Exercise	Limiting substance use 4 hours before sleep		
	30 min+[42]	Caffeine[43]	Nicotine[44]	Alcohol[45]
Sleep environment	Sensory deprivation		
	Eye mask and blackout curtains[46]	Ear plugs and white noise[47]	
Sleep time	Optimal sleep time	Regular sleep time	
	6–8 hours[48]	Sleep schedule[49]	

Sleeping, like exercise and fasting, also exists on a spectrum. Regarding sleep, HGH is secreted only after the stage 3 threshold is crossed. Getting into stage 3 sleep is not only a matter of the time it takes to enter that stage, but also the ability to minimize disturbances that may keep you from entering deeper stages of sleep or staying in that stage. You can sleep lightly for as many hours as you want, but you will not secrete HGH from sleep if you do not enter stage 3. Figure 2–13 displays a spectrum of the different stages of sleep. Stage 3 has the lowest brain activity and REM has the highest.

Sleep spectrum

The thick line on the far right of the chart in figure 2–13 represents consciousness. To the right of the far right vertical line would be where the GBA spectrum begins. Sleep stages (depth or intensity of sleep) are determined by brain waves, which can be measured using an electroencephalogram (EEG). As you get into deeper stages of sleep, your brain waves slow down. As you get closer to consciousness, your brain waves speed up.

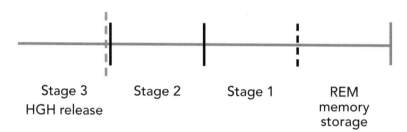

FIGURE 2–13. Sleep spectrum

Sleep cycles

Throughout the night, your brain attempts to complete natural sleep cycles, which take you from a conscious state into stage 1, then stage 2, then stage 3, and then to REM sleep. Figure 2–14 shows what a sleep cycle could look like if your brain is entering stage 3 sleep as it should under normal conditions. As the night goes on, the time spent in stage 3 sleep continues to decrease.

Firefighters know that sleeping conditions at the firehouse are not considered normal. Disrupted sleep while on duty is an accepted part of being a firefighter. Furthermore, sleep is an unconscious physical activity, which means you are not in control of your sleep stages. In order to maximize the HGH benefits of sleep, you need to ensure that you can get the best sleep with the least disruptions possible on the nights you are not working.

You may be skeptical of the idea that you can influence how deeply you sleep on your rest and recovery days. Additionally, because sleep is not a conscious activity, you may believe it should not be included in the activities that

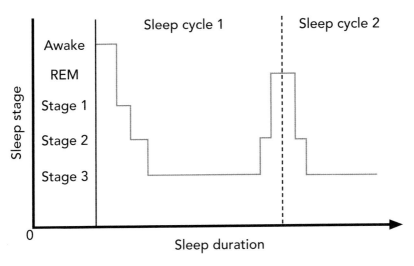

FIGURE 2–14. Sleep cycles

fall under your control. In that case, you might dismiss the sleep spectrum chart as irrelevant. If you are in the category just described, then consider the implications of a weekly sleep chart in figure 2–15.

> Fire department shift-work schedules are designed to allow for rest and recovery on days off. However, overtime, part-time jobs, and volunteering provide opportunities for firefighters to earn more money or gain more experience. The trade-off is often between the extra money and experience versus rest and recovery, including sleep.

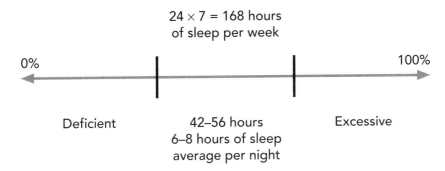

FIGURE 2–15. Weekly sleep spectrum

The suggested amount of sleep for the average person is 6–8 hours per night. In figure 2–15, you can see 6–8 hours of sleep each night means 42–56 hours of sleep per week. If you are sleeping for less than 42 hours or more than 56 hours each week, it may be worth considering whether the choices you are making regarding your sleep schedule are deficient or excessive. At the very least, you can control whether you choose to work overtime, part-time, or volunteer. If your decisions to make more money or volunteer come at the cost of sleep on your off-duty days, the fault for sleep-related health issues lies only in yourself.

The S Curve and Finding a Balance

Aristotle's concern with finding the mean had to do with understanding what actions were deficient, what actions were excessive, and then choosing something that was balanced. While Aristotle was concerned about philosophical matters such as virtue ethics, he was also very concerned about how deficiency and excess were related to practical matters. Aristotle understood each individual is different, and those differences can change what it means to be deficient or excessive regarding exercise, fasting, and sleep for each person.

Finding a balance does not mean always being in the middle of a spectrum of activity. This is impossible with regard to sleeping for instance, or even exercising. You cannot be in Gear 3 all the time and you should not be sleeping or sitting all day. Even during sleep, your brain cycles through stages of higher activity and lower activity throughout the night. We cannot stay in a fasting state forever, and we should not eat all day long. It is necessary at times to engage in different parts of the spectrum of activity, as long as the overall outcome is balanced.

You have the ability to push the relative limits of what you can do. This is possible because humans are adaptable. The ability to push the limits is very real, but for practical purposes, the limits become harder and harder to push as you get better at something. This is true for acute activity (one round of exercise being more intense than the last) and for long-term trends (overall ability to meet a goal). Figure 2–16 shows what is known as an S curve, which represents the idea that change has lower and upper thresholds.

The concept of the S curve is that change occurs at different speeds on the road to mastery.[50] Initially, change is slow until you cross a certain threshold. After that, rapid growth becomes possible and the payoff of effort is greater. Eventually, growth slows down and becomes incremental, and the payoff of effort becomes much less.

The S curve can be applied to anything that you can improve. For fasting, it could be applied to how long you fast or how many days each week you fast.

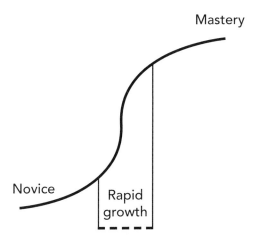

FIGURE 2–16. The S curve

For sleep it could be applied to improving your sleep hygiene practices, getting enough sleep each night, or getting enough sleep each week. At some point, you will have reached a level of relative mastery, and any more growth is incremental at best or excessive at worst. For exercise this is also true. However, there are many types of exercise, and each one has an S curve. Reaching mastery in running does not make you an expert at squatting.

The takeaway is that you can be deficient or excessive in either a single instance or a long-term trend. Once you have reached mastery for an activity, you can overdo it (be excessive) by trying to do too much. Instead, consider improving another S curve.

Assessing Your Physical Fitness Lifestyle

HGH can be optimized in adulthood through exercise, fasting, and sleep. For firefighters that work at a busy station, all three activities may be out of the question while on duty. However, if the goal is to optimize physical wellness, firefighters should seriously consider ways to incorporate exercise, fasting, and sleep hygiene whenever possible, even if that is only while off duty.

> Exercise, fasting, and sleep are ways to naturally increase HGH by 100s or even 1000s of percent above baseline level. Naturally occurring HGH does not have side effects and is a normal part of human health and wellness. However, there are side effects from using synthetic growth hormone or having hormone replacement therapy.

By naturally increasing HGH, you can increase the amount of time you spend as a healthy person. Remember that HGH naturally declines as you age, but you can increase the amount of HGH by exercising in Gear 2 or higher, fasting for 12 hours or longer, and sleeping well enough to enter stage 3 sleep (hopefully, multiple times each night). Firefighters are exposed to more stress than the average American, so they should make more effort to optimize wellness than the average American.

Over the course of a 24/48 shift cycle, there are nine opportunities to produce HGH. Figure 2–17 is a job aid that can be used for optimizing HGH. The purpose of the chart is to set a goal for exercise, fasting, and sleeping, and then simply check off whether you completed each goal on each shift day. If you are checking nine out of nine boxes with activities that go beyond the minimum threshold for HGH, then you are fully optimizing your HGH production.

It comes with the firefighting profession that night shifts may result in disrupted sleep. One night of disrupted sleep, and perhaps disrupted HGH, is not the end of the world. However, if you never exercise or fast and you work at a busy station, you may only be getting good HGH production during sleep on

Growth hormone daily chart (24/48)			
Shift	Gear 2+	Fasting	Sleeping
A shift			
B shift			
C shift			

HGH goals	What is your goal for each?
Gear 2+	
Fasting	
Sleeping	

FIGURE 2–17. HGH and 24/48 shift work

your off-duty days (two out of nine opportunities). And that is assuming you are not out partying on your nights off, working overtime, or volunteering somewhere else where your sleep may be disrupted. It is possible that some firefighters have zero out of nine opportunities for HGH on a regular shift cycle because of their lifestyle choices, which would cause accelerated decay and aging.

You should be able to produce HGH for at least five out of nine opportunities. That would require exercise during your work shift and your two off days, and good sleep on both off days. It would require no fasting at all. You would then be producing HGH on more than half of the opportunities available from exercise, fasting, and sleeping methods. Figure 2–18 offers an alternate HGH chart for those who have normal work weeks.

Growth hormone daily chart (24/48)																					
Sunday			Monday			Tuesday			Wednesday			Thursday			Friday			Saturday		
Gear 2+	Fasting	Sleeping	G2+	F	S	G2+	F	S	G2+	F	S	G2+	F	S	G2+	F	S	G2+	F	S

HGH goals	What does a 10 out of 10 look like for the following goals?
Gear 2+	
Fasting	
Sleeping	

FIGURE 2–18. HGH and a normal work week

Part 3: Micro and Macro

Exercise and Health Outcomes
Exercise and heart attack
Running has a strong correlation with heart health because running requires you to be in at least Gear 2, which releases HGH. Running is a great exercise for research because there is no downtime between exercises, and because it is easy to compare exertion levels between runners by using a runner's speed. Furthermore, lots of people run, which makes it easy to get data on a large number of participants.

> NFPA 1500 and NFPA 1583 have adopted the language from Occupational Safety and Health Administration's training mandate that fire departments shall educate their members on how to prevent illness, injury, and death. You will find this language in chapter 4 (section 4.3.1) and chapter 5 (section 5.1.1) of NFPA 1500 and in chapter 8 (section 8.1.1) of NFPA 1583.

One major study that included over 55,000 adults found that running significantly reduces the risk of death from cardiovascular causes. Compared to non-runners, those who ran had a 45% lower risk of cardiac mortality.[51] Amazingly, the reduction in cardiac mortality was consistent even for those who ran for just 30–60 minutes per week, which is just 5–10 minutes a day. Another study of 230,000 people found that running reduced cardiac mortality by 30%, and concluded similarly that you do not have to run for more than 50 minutes a week to benefit.[52]

Another study, with over 50,000 adults, considered the differences in running intensity. It turns out that exercising at a higher exertion level is more beneficial, which makes sense because the higher the exertion level, the more HGH you create. Runners who averaged a 7-minute mile were much less likely to need medication for hypertension (72% less), cholesterol (78% less), or diabetes (67% less) compared to those who ran a 10-minute mile or slower.[53]

Lastly, there may be something important about being in Gear 2 for a minimum of 10 minutes. A study out of the Stanford University School of Medicine found that after just 10 minutes of running, a detectable change can be found in 9,815 molecules in your blood.[54] The molecules that changed were linked to metabolism, immunity, oxidative stress, and cardiovascular function. Stanford is working on a patent to map the data from observing those molecules. Another study found that 10 minutes of high intensity exercise was observed to be a threshold for HGH secretion.[55]

The takeaway here is that there is abundant evidence that getting into Gear 2 or higher can help to protect your heart. Running is not the only way to get into Gear 2, but it is one of the easiest and cheapest ways. However, you can get into Gear 2 with any exercise you want to do, as long as the exertion level is high enough.

Exercise and cancer

In 2017, the Centers for Disease Control and Prevention (CDC) reported that 13 cancers are linked to being overweight or obese. Furthermore, the cancers associated with weight make up 40% of all cancers diagnosed.[56] The National

Institute for Occupational Safety and Health (NIOSH) firefighter cancer study included 19 cancers associated with firefighters; 8 of them overlap with the 13 associated with body weight.[57]

The 8 overlapping cancers from the NIOSH study and the CDC report include 4 cancers of the digestive system (esophagus, stomach, large intestine/colon, and rectum). In fact, the NIOSH study stated that they "observed excess digestive cancers mainly of the esophageal and colorectal sites." The authors went on to explain that lifestyle factors of diet, obesity, physical activity, tobacco, and alcohol are more likely than workplace hazards to be the cause of digestive cancers. In fact, the study of 230,000 people that found that running reduces cardiac mortality by 30% also found that running reduces cancer mortality by 23%, as long as you run for 50 minutes or more per week.[58] Again, running is just a proxy for exercise in Gear 2 or higher.

Available research calls out cancer as a mitochondrial metabolic disease.[59] You have probably been told that mitochondria are the "powerhouse of the cell." However, mitochondria do more than just make energy; they allow your cells to breathe (cellular respiration). If you suddenly lost your lungs, you would die fairly soon. If a cell suddenly lost its mitochondria, the cell would die fairly soon as well. And the best way to improve your mitochondrial respiration is to put your body under respiratory stress (i.e., exercise). With unhealthy or damaged mitochondria, your cells attempt to produce energy through fermentation, and this is one of the major pathways to cancer. Healthy mitochondria can actually suppress the reproduction of cancerous cells.

As it turns out, exercise plays a significant role in the health of your mitochondria. Exercise not only increases the ability of mitochondria to function, it also increases the number of mitochondria you have in your cells.[60] In addition, exercise helps your body regulate the replacement cycle of old or dying mitochondria, known as mitophagy.[61] Conversely, being overweight or obese has a negative impact on mitochondria health.[62] Perhaps that is why some researchers say that exercise is medicine for the mitochondria.[63] And, healthy cells make for a healthy body.

Exercise and suicide

There is an ongoing debate about mental disorders and suicide. Specifically, the debate focuses on the role of mental disorders in suicidal ideation and completed suicide. Data from around the world suggests that mental disorders play a major role in suicide.[64] As a reminder, psychological autopsies suggest that roughly 90% of people who completed suicide had treatable mental disorders, such as depression, anxiety, stress disorders, and addiction.[65] While having a mental disorder does not mean you will die by suicide, it does increase the risk for suicidal ideation and attempts.

A study of European adolescents found an interesting pattern relating to suicidal behavior.[66] The study compared individuals with lifestyles considered "high-risk" and "low-risk" to determine the risk of suicide. The high-risk group had a host of unhealthy behaviors, including drug and alcohol use, while the low-risk group had almost no unhealthy behaviors. Interestingly, an "invisible risk" group emerged from the data. The invisible risk group had three behaviors that made them virtually identical with the high-risk group, but whose behaviors are not stereotypically considered "high-risk." Those three behaviors were poor sleep, sedentary behavior, and high media use.

> There is a lot of research on firefighters having issues with sleep and sedentary behavior. There is currently not much research on firefighter media use, but if you are still active in the fire station, you know that media use and screen time is a problem.

As it turns out, exercise has a major impact on brain health and brain functionality. According to Dr. John Ratey, author of *Spark* and a psychiatrist and professor at Harvard, exercise is critical for the prevention and management of common mental disorders. Ratey explains that during exercise, your brain produces chemicals that not only help your brain grow, but also manage emotions. Chemicals such as dopamine, serotonin, gamma-aminobutyric acid (GABA), and brain-derived neurotropic factor (BDNF) all work to fight depression, anxiety, stress, and even addiction and attention deficit issues. In fact, Ratey shares the existing research on how exercise is better than depression and anxiety medications in the large majority of cases![67]

Exercise does more than just help us to feel better emotionally; it increases blood to the brain.[68] The higher the exercise intensity, the more blood flow to the brain.[69] One major study on brain dysfunction used brain scans on more than 17,000 individuals to measure cerebral blood flow.[70] Poor blood flow to the brain happens to be the number one predictor for Alzheimer's.[71] Poor blood flow to the brain is also associated with depression, ADHD, bipolar disorder, addiction, schizophrenia, and suicide. Moreover, the more fat you carry, the less blood your brain gets. Being sedentary and overweight are two separate ways to increase your risk of poor brain health.

Fasting and Health Outcomes
Fasting and heart attack
Fasting can be a preventive physical activity for lowering the risk of heart disease. Fasting has been shown to optimize body systems that affect the heart,

including lipid metabolism (cholesterol balancing), reduction of inflammation, reduction of body weight, reduction of blood pressure, reduction of resting heart rate, and optimization of glucose and insulin.[72] Fasting also optimizes cholesterol levels, including total cholesterol, triglycerides, and even specifically reducing harmful low-density lipoproteins (LDL) particles while increasing helpful LDL particles, which in turn lowers an individual's risk for heart disease.[73] In other words, fasting reduces the main risk factors of heart disease.

Studies are being done now to quantify the impact that fasting has on a person's life. One study found that in a group of 2,000 adults, those who had been routinely fasting for over five years had a 45% lower mortality rate than those adults who were not routinely fasting.[74] Another finding from the same study group, which was published in a separate research article, found that the routine fasters had a 71% lower rate of developing heart failure![75] It is important to remember that intermittent fasting and calorie restriction are not the same thing, and that intermittent fasting can provide health and longevity benefits without necessarily resulting in weight loss.[76]

Fasting and cancer
Fasting can be a preventive physical activity for lowering cancer risk. One of the main reasons might be because of the process called autophagy, which means self-eating. Dr. Yoshinori Ohsumi was awarded the Nobel Prize in Medicine or Physiology in 2016 for his lifetime of work on understanding autophagy.[77] Essentially, when an organism goes into a negative energy balance, a trigger causes the organism to find cells that are dying, weak, or even cancerous and devour those cells first for energy metabolism. So, by fasting and prompting the body to go into autophagy, you are literally instructing your body to devour cancer cells.

Fasting also prompts your body to increase stem cell regeneration. In 2014, University of Southern California researchers found that by doing prolonged fasts of 48–72 hours, the body can regenerate stem cells, which can help to repair and improve the immune system.[78] In 2018, a Massachusetts Institute of Technology study published a report that mice that fasted for just 24 hours had regeneration in the stem cells in their intestines.[79] Stem cell functionality and proper regeneration of cells, organs, and systems in the body is critical to staying healthy. Please note that this chapter is not recommending fasting for 48–72 hours, but fasting for 24 hours while off duty may be an easy way for firefighters to help prevent cancer.

Lastly, fasting improves mitochondrial functionality. During periods of fasting, mitochondria switch from using glucose as a primary fuel to using fat (ketones) as a primary fuel.[80] Most tumor cells cannot survive on fat for fuel; therefore, during a fasting state, the tumor cells either stop growing or begin dying. Fasting has also been shown to balance the reproduction and planned destruction of mitochondria, which leads to healthier function of both mitochondria and the cell.[81] Furthermore, Harvard T.H. Chan School of Public Health researchers found that under fasting conditions, mitochondria networks are enhanced by controlling the ability for mitochondria to fuse or split.[82] When the balance between mitochondria fusion and fission (splitting) gets thrown off, cancer can occur.

Fasting and suicide
Fasting can lower the risk of mental disorders. Similar to exercise, fasting can stimulate neurogenesis (the growth of new brain cells) by producing more of the chemical BDNF.[83] Dr. Ratey calls BDNF "miracle grow for the brain." BDNF allows for new neural connections to be made, repairing aging brain cells and protecting healthy cells, and is now believed to play a role in stress resistance and resilience.[84] Serotonin is another brain chemical that was historically believed to reduce depression, which is what selective serotonin reuptake inhibitors (SSRIs), such as Prozac, target. Interestingly, a major study published in 2022 looking at the role of serotonin in depression found that there may be no actual link there after all.[85] Whether serotonin does affect mood or not, it appears to be a contributor in many other areas of brain and body health.[86] Fasting appears to increase levels of serotonin, BDNF, and nerve growth factor (NGF).[87]

Other studies have found mental health benefits of fasting. Fasting increases the hunger hormone called ghrelin, which might act as a natural antidepressant.[88] Fasting seems to increase the reported subjective well-being of individuals.[89] Additionally, one study on nurses who fasted during Ramadan found that depression and stress were significantly reduced.[90]

The evidence is mounting that fasting is beneficial for your mental health. Fasting appears to optimize chemicals that allow your brain to grow, and it also appears to optimize your mood by increasing feelings of well-being and decreasing feelings of stress and depression. Therefore, consider fasting as a tool to add to your toolbox for reducing risks of suicide.

Sleep and Health Outcomes
Sleep and heart attack
Sleep quality is affected by your choices of activity during the day. There is research that suggests that fasting can improve sleep.[91] As previously mentioned, your sleep quality improves with daily exercise. Interestingly, there seems to be a two-way street between sleep and exercise, where poor sleep can actually lower your physical activity levels.[92] It turns out that sleep is also a major factor of heart health.

A meta-analysis of 11 studies that included over 1 million people found that those who slept less than 6 hours or more than 8 hours, on average, had a higher risk of dying from coronary artery disease or stroke.[93] Another study found that men who were in their 50s and who had less than 5 hours of sleep on average had double the risk of a cardiac event by their early 70s.[94] It appears that not enough sleep, or too much sleep, is associated with a host of heart disease risk factors including hypertension, obesity, diabetes, and metabolic syndrome.[95]

In a meta-analysis that included 15 studies and roughly 475,000 people, getting less than 5 hours a night of sleep was associated with a 48% increase in coronary heart disease (CHD).[96] Interestingly, sleeping more than 9 hours a night was associated with a 38% increase in CHD. A similar meta-analysis found that all-cause mortality was also associated with sleeping too little or too much.[97]

The disruption of HGH due to loss of sleep may play a role in the risk of cardiovascular disease. That is because HGH seems to have a positive impact on cardiovascular health.[98] The good news is that if you are in your early 20s, you may still be getting enough HGH even with sleep loss or disruption, but that benefit does not last forever.[99] Regarding long-duration sleep, it may be that the lack of physical activity during the day plays a role in the increased mortality risk. The studies on sleep and heart health are associations, and not an exact cause and effect. However, getting 6–8 hours of sleep on your off-duty days is not a lot to ask in order to boost your chances of health.

If you feel that you cannot possibly get the recommended 6–8 hours of sleep per night on average, there may be some hope. One study reviewed data on over 10,000 participants for an average of 15 years per person and found that short sleep duration on its own did not significantly raise the risk of heart disease the way that sleep disruption did.[100] However, those who had both short sleep duration and disturbed sleep together were the most affected, with a 55% increase risk of heart disease. This seems to suggest that if you cannot possibly get 6–8 hours of sleep on average, then you should be intensely focused on optimizing your sleep quality to make every hour of shut-eye counts.

Sleep and cancer

When it comes to sleep and cancer, it seems that quality is the key issue. In fact, one meta-analysis that looked at 10 studies with over 555,000 participants specifically showed that sleep duration, whether short or long, is not associated with cancer.[101] However, sleep disruption appears to be harmful to hormone activity. Researchers at Stanford University found that sleep disruption changes at least two hormones—cortisol and melatonin.[102] Cortisol is also known as the "stress hormone" and plays a critical role in arousal, especially in preparing for fight or flight. When cortisol is released too often at night because of sleep disruption, or if cortisol timing is completely changed in the case of overnight shift work, the immune system can suffer from being disrupted too often, which can then lead to cancer development and growth.

Melatonin is thought to have antioxidant properties that protect cells from damage. So, the disruption of melatonin release decreases the cell's defense systems. Specifically for women, the decrease in melatonin allows estrogen levels to rise, which may play a role in increased prevalence of breast cancer.

At least two studies have shown an association between night-shift nurses and higher rates of breast cancer.[103] Night-shift nurses had roughly a 30% higher chance of breast cancer. But night shift seems to affect men as well. In a 2012 study that spanned six years with over 3,100 men, researchers found that compared with men that never worked night shift, those who did had an increased risk of cancers, including lung, colon, bladder, prostate, rectal, pancreatic, and non-Hodgkin's lymphoma.[104] Specifically, prostate cancer risk was almost three times higher in men who worked night shift compared to men who never did. Since the study of 3,100 men in 2012, other studies have been done on night shift and prostate cancer, and it seems that for prostate cancer specifically, the risk is in long-term shift work of 5 years or more.[105]

In 2019, the International Agency for Research on Cancer (IARC) classified night-shift work as a probable carcinogen.[106] Hopefully this means that more studies will start to look at night-shift work and cancer, and perhaps specifically at firefighters. In the meantime, the best way to help your body is to ensure good, uninterrupted nighttime sleep on your off-duty days.

Sleep and suicide

Like heart health and cancer, sleep has an impact on brain health. In 2013, the American Psychological Association (APA) collected data on the relationship between sleep and stress, using a survey to identify how much sleep Americans say they get and how they feel. The survey results demonstrated a self-reported pattern that either less sleep or poor-quality sleep increased stress in adults.[107] Beyond survey data, an association appears to exist between sleep duration, sleep quality, and cortisol (stress hormone) levels. The less sleep you get, or the more disruptions during sleep, the more cortisol your body produces during waking hours.[108] This matters because, according to the APA study, you are more likely to feel angry, overwhelmed, and low energy; snap at your kids; and even skip exercise when you do not get enough good sleep. However, stress is not the only issue that comes from poor or disrupted sleep.

Poor sleep is also associated with a higher risk of suicide. Recall that the study that identified an "invisible risk" group of adolescents that were virtually identical to the high-risk group for suicide risk had sleep disturbance as one of the three main risk factors.[109] Other studies have also found a link between poor sleep and suicide.

In 2017, Stanford University researchers used electronic wrist devices to learn that sleep disturbance independently increases the risk of suicide symptoms.[110] The researchers found that the more change there was in the time that a participant went to bed and woke up, the higher the suicidal ideation risk. The results were the same even when the researchers controlled for depression, substance use, and suicidal symptoms at the start of the study.

In 2020, the results of a three-year study on sleep patterns and suicidal behavior were published. The study found that trouble falling asleep, early morning awakening, and excessive sleepiness were associated with an increased risk of suicidal behavior independent of existing mental disorders.[111] In other words, even though suicidal behavior is associated with mental disorders, the study was able to rule out the contribution of any mental disorders in this study. A body of literature has identified a link between deficient sleep and suicidal behavior.[112]

Conclusion

1. You cannot change your stress response system, natural HGH decline, or your body type.
2. You can optimize your HGH, and thus your health, through exercise, fasting, and sleep.
3. Optimizing your HGH requires:
   a. Exercising in Gear 2 or higher
   b. Fasting for 12 hours or longer
   c. Ensuring that you get enough uninterrupted stage 3 sleep (ideally 6–8 hours) on your off-duty shifts
4. Your choices for physical activity have a direct effect on your body systems, organs, and cells.
5. The health of your body systems, organs, and cells impacts your risk of heart attack, cancer, and suicide.

Heart attack, cancer, and suicide are health issues that are killing firefighters, more so than workplace trauma. Fire departments must take active measures to educate their members on how to optimize physical wellness. Firefighters must then make wellness choices that optimize their health. Until these two efforts are occurring, demands for support regarding heart attack, cancer, and suicide will not be taken completely seriously by legislators and the public.

Your body is designed to operate as efficiently as possible. The less physical activity you do, the more your body breaks itself down so it can be more cost effective. In other words, each calorie will go farther if there is less of your body to feed. Unfortunately, this means that your body can and will weaken its own body systems, organs, and cells (including the heart, brain, and muscles), if it sees no reason to keep building them stronger. Optimizing your HGH by controlling your physical activity will signal your body to stay healthy and reduce the risks of illness or death.

Questions

1. What are three things that you cannot change about your body when it comes to physical fitness?
2. What are three things that you can change about your body when it comes to physical fitness?
3. What is the equation for exercise volume?
4. How many METs do you burn while sitting?
5. The measurement of 1 MET is your exertion level while at rest (e.g., sitting). (True/False)
6. How many METs is the minimum to qualify for vigorous exercise?
7. According to NFPA 1582, how many METs is required for the annual firefighter Bruce protocol stress test?
8. In which Gear does HGH begin to be released during exercise?
9. What are the two main categories of intermittent fasting?
10. What are the different types within the two main categories of intermittent fasting?
11. How many hours of fasting do you need before HGH begins to be released?
12. What are the four different stages of sleep?
13. In which stage does HGH get released?
14. What is the optimal amount of sleep for a regular night? For a regular week?
15. What is the S curve?
16. Give one example for how exercise reduces your risk of heart attack, cancer, and suicide.
17. Give one example for how fasting reduces your risk of heart attack, cancer, and suicide.
18. Give one example for how sleep reduces your risk of heart attack, cancer, and suicide.

19. Which chapters in NFPA 1500 relate to physical fitness and general wellness?
20. Which chapters in NFPA 1583 relate to physical fitness and health promotion education?

Notes

1. "Three-Quarters of Workers Are Stressed, Says New CareerCast Survey," *Cision PR Newswire*, March 6, 2019, https://www.prnewswire.com/news-releases/three-quarters-of-workers-are-stressed-says-new-careercast-survey-300807080.html; Simone Johnson, "The Top 10 Most and Least Stressful Jobs," *Business News Daily*, February 21, 2023, https://www.businessnewsdaily.com/1875-stressful-careers.html.
2. Patricia Watson et al., *Stress First Aid for Firefighters and Emergency Medical Services Personnel: Student Manual* (National Fallen Firefighters Foundation, 2013).
3. Chris Crowley and Henry S. Lodge, *Younger Next Year: Love Strong, Fit, and Sexy—Until You're 80 and Beyond* (Workman Publishing Company, 2007).
4. David E. Cummings and George R. Merriam, "Age-Related Changes in Growth Hormone Secretion: Should the Somatopause Be Treated?" *Seminars in Reproductive Endocrinology* 17, no. 4 (1999): 311–25, https://doi.org/10.1055/s-2007-1016241.
5. William Herbert Sheldon, Stanley Smith Stevens, and William Boose Tucker, *The Varieties of Human Physique* (Harper, 1940).
6. Jessie Newell, "Anthropometric Measurements: When to Use This Assessment," *ACE*, April 11, 2014, https://www.acefitness.org/fitness-certifications/ace-answers/exam-preparation-blog/3815/anthropometric-measurements-when-to-use-this-assessment/.
7. Carlos Celis-Morales et al., "Physical Activity Attenuates the Effect of the FTO Genotype on Obesity Traits in European Adults: The Food4Me Study," *Obesity* 24, no. 4 (2016), 962–69, https://doi.org/10.1002/oby.21422.
8. Walker S. C. Poston et al., "The Prevalence of Overweight, Obesity, and Substandard Fitness in a Population-Based Firefighter Cohort," *Journal of Occupational and Environmental Medicine* 53, no. 3 (2011): 266–73, https://doi.org/10.1097/JOM.0b013e31820af362.
9. Carolyn M. Reyes-Guzman et al., "Overweight and Obesity Trends Among Active Duty Military Personnel: A 13-Year Perspective," *American Journal of Preventive Medicine* 48, no. 2 (2014), 145–53, https://doi.org/10.1016/j.amepre.2014.08.033; Tracey J. Smith et al., "Overweight and Obesity in Military Personnel: Sociodemographic Predictors," *Obesity* 20, no. 7 (2012): 1534–38, https://doi.org/10.1038/oby.2012.25.
10. Jody L. Clasey et al., "Abdominal Visceral Fat and Fasting Insulin Are Important Predictors of 24-Hour GH Release Independent of Age, Gender, and

Other Physiological Factors," *The Journal of Clinical Endocrinology & Metabolism* 86, no. 8 (2001): 3845–52; Nicoleta Cristina Olarescu et al., "Normal Physiology of Growth Hormone in Adults," *Endotext* (2000), https://www.ncbi.nlm.nih.gov/books/NBK279056/.

11. Mohammed S. Ellulu et al., "Obesity and Inflammation: The Linking Mechanism and the Complications," *Archives of Medical Science* 13, no. 4 (2017): 851–63, https://doi.org/10.5114/aoms.2016.58928.
12. Aakash K. Patel et al., "Physiology, Sleep Stages," *StatPearls* (2022), https://www.ncbi.nlm.nih.gov/books/NBK526132/.
13. Nancy E. Felsing, Jo Anne Brasel, and Dan M. Cooper, "Effect of Low and High Intensity Exercise on Circulating Growth Hormone in Men," *The Journal of Clinical Endocrinology & Metabolism* 75, no. 1 (1992): 157–62, https://doi.org/10.1210/jcem.75.1.1619005; Cathy J. Pritzlaff et al., "Impact of Acute Exercise Intensity on Pulsatile Growth Hormone Release in Men," *Journal of Applied Physiology* 87, no. 2 (1999): 498–504, https://doi.org/10.1152/jappl.1999.87.2.498.
14. Bob Murray and Christine Rosenbloom, "Fundamentals of Glycogen Metabolism for Coaches and Athletes," *Nutrition Reviews* 76, no. 4 (2018): 243–59, https://doi.org/10.1093/nutrit/nuy001; Jørgen Jensen et al., "The Role of Skeletal Muscle Glycogen Breakdown for Regulation of Insulin Sensitivity by Exercise," *Frontiers in Physiology* 2, no. 112 (2011): 112, https://doi.org/10.3389/fphys.2011.00112.
15. Crowley and Lodge, *Younger Next Year.*
16. Julien S. Baker, Marie Clare McCormick, and Robert A. Roberts, "Interaction Among Skeletal Muscle Metabolic Energy Systems During Intense Exercise," *Journal of Nutrition and Metabolism* 2010 (2010): 905612, https://doi.org/10.1155/2010/905612; Alan R. Morton, "Exercise Physiology," in *Pediatric Respiratory Medicine*, 2nd ed., ed. Lynn M. Taussig and Louis I. Landau (Mosby, 2008), 89–99.
17. Philipp Mergenthaler et al., "Sugar for the Brain: The Role of Glucose in Physiological and Pathological Brain Function," *Trends in Neurosciences* 36, no. 10 (2013): 587–97, https://doi.org/10.1016/j.tins.2013.07.001.
18. Ken H. Darzy et al., "The Impact of Short-Term Fasting on the Dynamics of 24-Hour Growth Hormone (GH) Secretion in Patients with Severe Radiation-Induced GH Deficiency," *The Journal of Clinical Endocrinology & Metabolism* 91, no. 3 (2006): 987–94, https://doi.org/10.1210/jc.2005-2145; Klan Y. Ho et al., "Fasting Enhances Growth Hormone Secretion and Amplifies the Complex Rhythms of Growth Hormone Secretion in Man," *The Journal of Clinical Investigation* 81, no. 4 (1988): 968–75, https://doi.org/10.1172/JCI113450.
19. Sheldon, Stevens, and Tucker, *Varieties of Human Physique.*
20. "Frequency, Duration, and Intensity," DAPA Measurement Toolkit, accessed July 14, 2023, https://www.measurement-toolkit.org/physical-activity/introduction/frequency-duration-and-intensity.
21. Masahiro Banno et al., "Exercise Can Improve Sleep Quality: A Systematic Review and Meta-analysis," *PeerJ* 6 (2018): e5172, https://doi.org/10.7717/peerj.5172.

22. *NFPA 1582: Standard on Comprehensive Occupational Medical Program for Fire Departments* (National Fire Protection Association, 2022).
23. "Know Your Mets," Firefighter Safety Through Advanced Research, accessed July 14, 2023, https://www.fstaresearch.org/infographics-directory/?AnyType=true&FstarType=Infographic&PageSize=15.
24. Pritzlaff et al., "Impact of Acute Exercise Intensity."
25. Felsing, Brasel, and Cooper, "Effect of Low and High Intensity Exercise."
26. Aleksander S. Popel and Paul C. Johnson, "Microcirculation and Hemorheology," *Annual Review of Fluid Mechanics* 37 (2005): 43–69, https://doi.org/10.1146/annurev.fluid.37.042604.133933.
27. Kenneth R. Feingold, "Introduction to Lipids and Lipoproteins," *Endotext*, June 10, 2015.
28. Stephen D. Anton et al., "Flipping the Metabolic Switch: Understanding and Applying the Health Benefits of Fasting," *Obesity* 26, no. 2 (2018): 254–68, https://doi.org/10.1002/oby.22065.
29. Ho et al., "Fasting Enhances Growth Hormone Secretion."
30. "Ramadan 2021: Fasting Hours Around the World," *Aljazeera*, April 7, 2021, https://www.aljazeera.com/news/2021/4/7/ramadan-2021-fasting-hours-around-the-world.
31. Ho et al., "Fasting Enhances Growth Hormone Secretion"; "New Research Finds Routine Periodic Fasting Is Good for Your Health, and Your Heart," Intermountain Healthcare, last modified April 3, 2011, https://intermountainhealthcare.org/news/2011/04/new-research-finds-routine-periodic-fasting-is-good-for-your-health-and-your-heart/.
32. Darzy et al., "The Impact of Short-Term Fasting."
33. Humaira Jamshed et al., "Early Time-Restricted Feeding Improves 24-Hour Glucose Levels and Affects Markers of the Circadian Clock, Aging, and Autophagy in Humans," *Nutrients* 11, no. 6 (2019): 1234, https://doi.org/10.3390/nu11061234.
34. Chul Won Yun and Sang Hun Lee, "The Roles of Autophagy in Cancer," *International Journal of Molecular Sciences* 19, no. 11 (2018): 3466, https://doi.org/10.3390/ijms19113466.
35. "Stress and Sleep," American Psychological Association, last modified January 1, 2013, https://www.apa.org/news/press/releases/stress/2013/sleep; Camila Hirotsu, Sergio Tufik, and Monica Levy Andersen, "Interactions Between Sleep, Stress, and Metabolism: From Physiological to Pathological Conditions," *Sleep Science* 8, no. 3 (2015): 143–52, https://doi.org/10.1016/j.slsci.2015.09.002.
36. Xiaoli Shen, Yili Wu, and Dongfeng Zhang, "Nighttime Sleep Duration, 24-Hour Sleep Duration and Risk of All-Cause Mortality Among Adults: A Meta-analysis of Prospective Cohort Studies," *Scientific Reports* 6, no. 1 (2016): 21480, https://doi.org/10.1038/srep21480.
37. Jane E. Ferrie et al., "A Prospective Study of Change in Sleep Duration: Associations with Mortality in the Whitehall II Cohort," *Sleep* 30, no. 12 (2007): 1659–66, https://doi.org/10.1093/sleep/30.12.1659; Francesco P. Cappuccio et al.,

"Sleep Duration and All-Cause Mortality: A Systematic Review and Meta-Analysis of Prospective Studies," *Sleep* 33, no. 5 (2010): 585–92, https://doi.org/10.1093/sleep/33.5.585.
38. Eric Suni and Nilong Vyas, "Stages of Sleep," Sleep Foundation, last modified August 8, 2023, https://www.sleepfoundation.org/stages-of-sleep.
39. Y. Takahashi, D. M. Kipnis, and W. H. Daughaday, "Growth Hormone Secretion During Sleep," *The Journal of Clinical Investigation* 47, no. 9 (1968): 2079–90, https://doi.org/10.1172/JCI105893; Eve Van Cauter, Rachel Leproult, and Laurence Plat, "Age-Related Changes in Slow Wave Sleep and REM Sleep and Relationship with Growth Hormone and Cortisol Levels in Healthy Men," *JAMA* 284, no. 7 (2000): 861–68, https://doi.org/10.1001/jama.284.7.861.
40. Björn Rasch and Jan Born, "About Sleep's Role in Memory," *Physiological Reviews* 93, no. 2 (2013): 681–766, https://doi.org/10.1152/physrev.00032.2012; National Institutes of Health, "REM Sleep May Help the Brain Forget," NIH Research Matters, last modified October 1, 2019, https://www.nih.gov/news-events/nih-research-matters/rem-sleep-may-help-brain-forget.
41. Brett A. Dolezal et al., "Interrelationship Between Sleep and Exercise: A Systematic Review," *Advances in Preventive Medicine* 2017 (2017): 1364387, https://doi.org/10.1155/2017/1364387; Banno et al., "Exercise Can Improve Sleep Quality"; Kathrin Wunsch, Nadine Kasten, and Reinhard Fuchs, "The Effect of Physical Activity on Sleep Quality, Well-Being, and Affect in Academic Stress Periods," *Nature and Science of Sleep* (2017): 117–26, https://doi.org/10.2147/NSS.S132078; Frederick Baekeland and Richard Lasky, "Exercise and Sleep Patterns in College Athletes," *Perceptual and Motor Skills* 23, no. 3 Suppl. (1966): 1203–7, https://doi.org/10.2466/pms.1966.23.3f.1203.
42. Dolezal et al., "Interrelationship Between Sleep and Exercise"; Banno et al., "Exercise Can Improve Sleep Quality"; Wunsch, Kasten, and Fuchs, "The Effect of Physical Activity on Sleep Quality"; Baekeland and Lasky, "Exercise and Sleep Patterns"; "Exercising for Better Sleep," Johns Hopkins Medicine, accessed August 15, 2023, https://www.hopkinsmedicine.org/health/wellness-and-prevention/exercising-for-better-sleep.
43. S. H. Onen et al., "[Prevention and Treatment of Sleep Disorders through Regulation of Sleeping Habits]," *Presse Medicale* 23, no. 10 (1994): 485–89; Christine E. Spadola et al., "Evening Intake of Alcohol, Caffeine, and Nicotine: Night-to-Night Associations with Sleep Duration and Continuity among African Americans in the Jackson Heart Sleep Study," *Sleep* 42, no. 11 (2019): zsz136, https://doi.org/10.1093/sleep/zsz136; Lotan Shilo et al., "The Effects of Coffee Consumption on Sleep and Melatonin Secretion," *Sleep Medicine* 3, no. 3 (2002): 271–73; Christopher Drake et al., "Caffeine Effects on Sleep Taken 0, 3, or 6 Hours before Going to Bed," *Journal of Clinical Sleep Medicine* 9, no. 11 (2013): 1195–200, https://doi.org/10.5664/jcsm.3170; Ismet Karacan et al., "Dose-Related Sleep Disturbances Induced by Coffee and Caffeine," *Clinical Pharmacology & Therapeutics* 20, no. 6 (1976): 682–89.

44. Onen et al., "[Prevention and Treatment of Sleep Disorders]"; Andreas Jaehne et al., "Effects of Nicotine on Sleep During Consumption, Withdrawal and Replacement Therapy," *Sleep Medicine Reviews* 13, no. 5 (2009): 363–77, https://doi.org/10.1016/j.smrv.2008.12.003; Spadola et al., "Evening Intake of Alcohol, Caffeine, and Nicotine."
45. Onen et al., "[Prevention and Treatment of Sleep Disorders]"; Spadola et al., "Evening Intake of Alcohol, Caffeine, and Nicotine."
46. Joshua J. Gooley et al., "Exposure to Room Light before Bedtime Suppresses Melatonin Onset and Shortens Melatonin Duration in Humans," *The Journal of Clinical Endocrinology & Metabolism* 96, no. 3 (2011): E463–72, https://doi.org/10.1210/jc.2010-2098; Jounhong Ryan Cho et al., "Let There Be No Light: The Effect of Bedside Light on Sleep Quality and Background Electroencephalographic Rhythms," *Sleep Medicine* 14, no. 12 (2013): 1422–25, http://doi.org/10.1016/j.sleep.2013.09.007; Rong-fang Hu et al., "Effects of Earplugs and Eye Masks on Nocturnal Sleep, Melatonin and Cortisol in a Simulated Intensive Care Unit Environment," *Critical Care* 14, no. 2 (2010): R66, https://doi.org/10.1186/cc8965.
47. Hu et al., "Effects of Earplugs and Eye Masks"; Pouya Farokhnezhad Afshar et al., "Effect of White Noise on Sleep in Patients Admitted to a Coronary Care," *Journal of Caring Sciences* 5, no. 2 (2016): 103–9, https://doi.org/10.15171/jcs.2016.011; Kenneth I. Hume, Mark Brink, and Mathias Basner, "Effects of Environmental Noise on Sleep," *Noise and Health* 14, no. 61 (2012): 297–302, https://doi.org/10.4103/1463-1741.104897.
48. Shen, Wu, and Zhang, "Nighttime Sleep Duration"; Ferrie et al., "A Prospective Study of Change in Sleep Duration;" Cappuccio et al., "Sleep Duration and All-Cause Mortality."
49. Jiunn-Horng Kang, and Shih-Ching Chen, "Effects of an Irregular Bedtime Schedule on Sleep Quality, Daytime Sleepiness, and Fatigue among University Students in Taiwan," *BMC Public Health* 9, no. 1 (2009): Article 248, https://doi.org/10.1186/1471-2458-9-248; Kana Okano et al., "Sleep Quality, Duration, and Consistency Are Associated with Better Academic Performance in College Students," *NPJ Science of Learning* 4, no. 1 (2019): 16, https://doi.org/10.1038/s41539-019-0055-z; Eric Suni and Nilong Vyas, "Stages of Sleep," Sleep Foundation, accessed August 8, 2023, https://www.sleepfoundation.org/stages-of-sleep; Jean-Philippe Chaput et al., "Sleep Timing, Sleep Consistency, and Health in Adults: A Systematic Review," *Applied Physiology, Nutrition, and Metabolism* 45, no. 10 (2020): S232–47, https://doi.org/10.1139/apnm-2020-0032.
50. Nick van Dam, *25 Best Practices in Learning & Talent Development*, 2nd ed. (Lulu.com, 2007).
51. Duck-Chul Lee et al., "Leisure-Time Running Reduces All-Cause and Cardiovascular Mortality Risk," *Journal of the American College of Cardiology* 64, no. 5 (2014): 472–81, https://doi.org/10.1016/j.jacc.2014.04.058.

52. Zeljko Pedisic et al., "Is Running Associated with a Lower Risk of All-Cause, Cardiovascular and Cancer Mortality, and Is the More the Better? A Systematic Review and Meta-analysis," *British Journal of Sports Medicine* 54, no. 15 (2020): 898–905, https://doi.org/10.1136/bjsports-2018-100493.
53. Paul T. Williams, "Relationship of Running Intensity to Hypertension, Hypercholesterolemia, and Diabetes," *Medicine and Science in Sports and Exercise* 40, no. 10 (2008): 1740, https://doi.org/10.1249/MSS.0b013e31817b8ed1.
54. Kévin Contrepois et al., "Molecular Choreography of Acute Exercise," *Cell* 181, no. 5 (2020): 1112–30, https://doi.org/10.1016/j.cell.2020.04.043; Hanae Armitage, "Stanford Medicine Study Details Molecular Effects of Exercise," *Stanford Medicine News Center*, May 28, 2020, https://med.stanford.edu/news/all-news/2020/05/stanford-medicine-study-details-molecular-effects-of-exercise.html.
55. Felsing, Brasel, and Cooper, "Effects of Low and High Intensity Exercise."
56. "Cancers Associated with Overweight and Obesity Make up 40 Percent of Cancers Diagnosed in the United States," CDC Newsroom, last modified October 3, 2017, https://www.cdc.gov/media/releases/2017/p1003-vs-cancer-obesity.html.
57. Robert D. Daniels et al., "Mortality and Cancer Incidence in a Pooled Cohort of US Firefighters from San Francisco, Chicago and Philadelphia (1950–2009)," *Occupational and Environmental Medicine* 71, no. 6 (2014): 388–97, https://doi.org/10.1136/oemed-2013-101662.
58. Pedisic et al., "Is Running Associated with a Lower Risk."
59. Thomas N. Seyfried, "Cancer as a Mitochondrial Metabolic Disease," *Frontiers in Cell and Developmental Biology* 3 (2015): 43, https://doi.org/10.3389/fcell.2015.00043.
60. Elizabeth V. Menshikova et al., "Effects of Exercise on Mitochondrial Content and Function in Aging Human Skeletal Muscle," *The Journals of Gerontology Series A: Biological Sciences and Medical Sciences* 61, no. 6 (2006): 534–40, https://doi.org/10.1093/gerona/61.6.534.
61. Yuntian Guan, Joshua C. Drake, and Zhen Yan, "Exercise-Induced Mitophagy in Skeletal Muscle and Heart," *Exercise and Sport Sciences Reviews* 47, no. 3 (2019): 151–56, https://doi.org/10.1249/JES.0000000000000192; Estelle Balan et al., "Regular Endurance Exercise Promotes Fission, Mitophagy, and Oxidative Phosphorylation in Human Skeletal Muscle Independently of Age," *Frontiers in Physiology* 10 (2019): 1088, https://doi.org/10.3389/fphys.2019.01088.
62. Juan C. Bournat and Chester W. Brown, "Mitochondrial Dysfunction in Obesity," *Current Opinion in Endocrinology, Diabetes, and Obesity* 17, no. 5 (2010): 446–52, https://doi.org/10.1097/MED.0b013e32833c3026; Mary Madeline Rogge, "The Role of Impaired Mitochondrial Lipid Oxidation in Obesity," *Biological Research for Nursing* 10, no. 4 (2009): 356–73, https://doi.org/10.1177/1099800408329408; Aline Haas de Mello et al., "Mitochondrial Dysfunction in Obesity," *Life Sciences* 192 (2018): 26–32, https://doi.org/10.1016/j.lfs.2017.11.019.

63. Ashley N. Oliveira et al., "Exercise Is Muscle Mitochondrial Medicine," *Exercise and Sport Sciences Reviews* 49, no. 2 (2021): 67–76, https://doi.org/10.1249/JES.0000000000000250; Daniela Sorriento, Eugenio Di Vaia, and Guido Iaccarino, "Physical Exercise: A Novel Tool to Protect Mitochondrial Health," *Frontiers in Physiology* 12 (2021): 660068, https://doi.org/10.3389/fphys.2021.660068.
64. José Manoel Bertolote and Alexandra Fleischmann, "Suicide and Psychiatric Diagnosis: A Worldwide Perspective," *World Psychiatry* 1, no. 3 (2002): 181–85, https://www.ncbi.nlm.nih.gov/pmc/articles/PMC1489848/.
65. Louise Brådvik, "Suicide Risk and Mental Disorders," *International Journal of Environmental Research and Public Health* 15, no. 9 (2018): 2028, https://doi.org/10.3390/ijerph15092028.
66. Vladimir Carli et al., "A Newly Identified Group of Adolescents at 'Invisible' Risk for Psychopathology and Suicidal Behavior: Findings from the SEYLE Study," *World Psychiatry* 13, no. 1 (2014): 78–86, https://doi.org/10.1002/wps.20088.
67. John J. Ratey and Eric Hagerman, *Spark: The Revolutionary New Science of Exercise and the Brain* (Little, Brown Spark, 2013)
68. Jordi P. D. Kleinloog et al., "Aerobic Exercise Training Improves Cerebral Blood Flow and Executive Function: A Randomized, Controlled Cross-Over Trial in Sedentary Older Men," *Frontiers in Aging Neuroscience* 11 (2019): 333, https://doi.org/10.3389/fnagi.2019.00333; Jordan S. Querido and A. William Sheel, "Regulation of Cerebral Blood Flow During Exercise," *Sports Medicine* 37 (2007): 765–82, https://doi.org/10.2165/00007256-200737090-00002.
69. Shigehiko Ogoh and Philip N. Ainslie, "Cerebral Blood Flow During Exercise: Mechanisms of Regulation," *Journal of Applied Physiology* 107, no. 5 (2009): 1370–80, https://doi.org/10.1152/japplphysiol.00573.2009.
70. Daniel G. Amen et al., "Patterns of Regional Cerebral Blood Flow as a Function of Obesity in Adults," *Journal of Alzheimer's Disease* 77, no. 3 (2020): 1331–37, https://doi.org/10.3233/JAD-200655.
71. Chelsea C. Hays, Zvinka Z. Zlatar, and Christina E. Wierenga, "The Utility of Cerebral Blood Flow as a Biomarker of Preclinical Alzheimer's Disease," *Cellular and Molecular Neurobiology* 36 (2016): 167–79, https://doi.org/10.1007/s10571-015-0261-z; IOS Press, "Body Weight Has Surprising, Alarming Impact on Brain Function," *Science Daily*, August 5, 2020, https://www.sciencedaily.com/releases/2020/08/200805110127.htm.
72. Bartosz Malinowski et al., "Intermittent Fasting in Cardiovascular Disorders—An Overview," *Nutrients* 11, no. 3 (2019): 673, https://doi.org/10.3390/nu11030673.
73. Surabhi Bhutani et al., "Alternate Day Fasting and Endurance Exercise Combine to Reduce Body Weight and Favorably Alter Plasma Lipids in Obese Humans," *Obesity* 21, no. 7 (2013): 1370–79, https://doi.org/10.1002/oby.20353; Surabhi Bhutani et al., "Improvements in Coronary Heart Disease Risk Indicators by

Alternate-Day Fasting Involve Adipose Tissue Modulations," *Obesity* 18, no. 11 (2010): 2152–59, https://doi.org/10.1038/oby.2010.54; Haiyan Meng et al., "Effects of Intermittent Fasting and Energy-Restricted Diets on Lipid Profile: A Systematic Review and Meta-analysis," *Nutrition* 77 (2020): 110801, https://doi.org/10.1016/j.nut.2020.110801.

74. "Regular Fasting Could Lead to Longer, Healthier Life," American Heart Association, last modified November 25, 2019, https://www.heart.org/en/news/2019/11/25/regular-fasting-could-lead-to-longer-healthier-life; Benjamin D. Horne et al., "Intermittent Fasting Lifestyle and Human Longevity in Cardiac Catheterization Populations," *Circulation* 140, Suppl. 1 (2019): A11123.

75. "Regular Fasting"; Ciera Bartholomew et al., "Intermittent Fasting Lifestyle and Incidence of Heart Failure and Myocardial Infarction in Cardiac Catheterization Patients," *Circulation* 140, no. Suppl. 1 (2019): A10043.

76. Dae-Sung Hwangbo et al., "Mechanisms of Lifespan Regulation by Calorie Restriction and Intermittent Fasting in Model Organisms," *Nutrients* 12, no. 4 (2020): 1194, https://doi.org/10.3390/nu12041194.

77. "The Nobel Prize in Physiology or Medicine 2016: Yoshinori Ohsumi," The Nobel Prize, last modified October 3, 2016, https://www.nobelprize.org/prizes/medicine/2016/press-release/.

78. Chia-Wei Cheng et al., "Prolonged Fasting Reduces IGF-1/PKA to Promote Hematopoietic-Stem-Cell-Based Regeneration and Reverse Immunosuppression," *Cell Stem Cell* 14, no. 6 (2014): 810–23, https://doi.org/10.1016/j.stem.2014.04.014.

79. Maria M. Mihaylova et al., "Fasting Activates Fatty Acid Oxidation to Enhance Intestinal Stem Cell Function During Homeostasis and Aging," *Cell Stem Cell* 22, no. 5 (2018): 769–78, https://doi.org/10.1016/j.stem.2018.04.001.

80. Thomas N. Seyfried et al., "Cancer as a Metabolic Disease: Implications for Novel Therapeutics," *Carcinogenesis* 35, no. 3 (2014): 515–27, https://doi.org/10.1093/carcin/bgt480.

81. Mohammad Bagherniya et al., "The Effect of Fasting or Calorie Restriction on Autophagy Induction: A Review of the Literature," *Ageing Research Reviews* 47 (2018): 183–97, https://doi.org/10.1016/j.arr.2018.08.004; Heather J. Weir et al., "Dietary Restriction and AMPK Increase Lifespan via Mitochondrial Network and Peroxisome Remodeling," *Cell Metabolism* 26, no. 6 (2017): 884–96, https://doi.org/10.1016/j.cmet.2017.09.024.

82. Weir et al., "Dietary Restriction and AMPK."

83. Karin Seidler and Michelle Barrow, "Intermittent Fasting and Cognitive Performance–Targeting BDNF as Potential Strategy to Optimise Brain Health," *Frontiers in Neuroendocrinology* 65 (2022): 100971, https://doi.org/10.1016/j.yfrne.2021.100971.

84. Ratey and Hagerman, *Spark: The Revolutionary New Science of Exercise and the Brain.*

85. Jeffrey Fessel, "Formulating Treatment of Major Psychiatric Disorders: Algorithm Targets the Dominantly Affected Brain Cell-Types," *Discover Mental Health* 3, no. 1 (2023): 3, https://doi.org/10.1007/s44192-022-00029-8.

86. Miles Berger, John A. Gray, and Bryan L. Roth, "The Expanded Biology of Serotonin," *Annual Review of Medicine* 60 (2009): 355–66, https://doi.org/10.1146/annurev.med.60.042307.110802.
87. Abdolhossein Bastani, Sadegh Rajabi, and Fatemeh Kianimarkani, "The Effects of Fasting During Ramadan on the Concentration of Serotonin, Dopamine, Brain-Derived Neurotrophic Factor and Nerve Growth Factor," *Neurology International* 9, no. 2 (2017): 7043, https://doi.org/10.4081/ni.2017.7043.
88. Anjana Bali and Amteshwar Singh Jaggi, "An Integrative Review on Role and Mechanisms of Ghrelin in Stress, Anxiety and Depression," *Current Drug Targets* 17, no. 5 (2016): 495–507, https://doi.org/10.2174/1389450116666150518095650.
89. Mark P. Mattson et al., "Intermittent Metabolic Switching, Neuroplasticity and Brain Health," *Nature Reviews Neuroscience* 19, no. 2 (2018): 81–94, https://doi.org/10.1038/nrn.2017.156; Elisa Berthelot et al., "Fasting Interventions for Stress, Anxiety and Depressive Symptoms: A Systematic Review and Meta-Analysis," *Nutrients* 13, no. 11 (2021): 3947, https://doi.org/10.3390/nu13113947.
90. Ali Noruzi Koushali et al., "Effect of Ramadan Fasting on Emotional Reactions in Nurses," *Iranian Journal of Nursing and Midwifery Research* 18, no. 3 (2013): 232–36, https://journals.lww.com/jnmr/Fulltext/2013/18030/Effect_of_Ramadan_fasting_on_emotional_reactions.12.aspx.
91. Andreas Michalsen et al., "Effects of Short-Term Modified Fasting on Sleep Patterns and Daytime Vigilance in Non-Obese Subjects: Results of a Pilot Study," *Annals of Nutrition and Metabolism* 47, no. 5 (2003): 194–200, https://doi.org/10.1159/000070485.
92. Christopher E. Kline, "The Bidirectional Relationship Between Exercise and Sleep: Implications for Exercise Adherence and Sleep Improvement," *American Journal of Lifestyle Medicine* 8, no. 6 (2014): 375–79, https://doi.org/10.1177/1559827614544437.
93. Chayakrit Krittanawong et al., "Association Between Short and Long Sleep Durations and Cardiovascular Outcomes: A Systematic Review and Meta-Analysis," *European Heart Journal: Acute Cardiovascular Care* 8, no. 8 (2019): 762–70, https://doi.org/10.1177/2048872617741733.
94. M. Bengtsson et al., "Middle Age Men with Short Sleep Duration Have Two Times Higher Risk of Cardiovascular Events Than Those with Normal Sleep Duration, a Cohort Study with 21 Years Follow-Up," *European Heart Journal* 39, Suppl. 1 (2018): ehy565.P2465, https://doi.org/10.1093/eurheartj/ehy565.P2465.
95. Bo Xi et al., "Short Sleep Duration Predicts Risk of Metabolic Syndrome: A Systematic Review and Meta-analysis," *Sleep Medicine Reviews* 18, no. 4 (2014): 293–97, https://doi.org/10.1016/j.smrv.2013.06.001.
96. Francesco P. Cappuccio et al., "Sleep Duration Predicts Cardiovascular Outcomes: A Systematic Review and Meta-Analysis of Prospective Studies," *European Heart Journal* 32, no. 12 (2011): 1484–92, https://doi.org/10.1093/eurheartj/ehr007.
97. Cappuccio et al., "Sleep Duration and All-Cause Mortality."

98. Diego Caicedo et al., "Growth Hormone (GH) and Cardiovascular System," *International Journal of Molecular Sciences* 19, no. 1 (2018): 290, https://doi.org/10.3390/ijms19010290.

99. Janet Mullington et al., "Age-Dependent Suppression Of Nocturnal Growth Hormone Levels During Sleep Deprivation," *Neuroendocrinology* 64, no. 3 (1996): 233–41, https://doi.org/10.1159/000127122.

100. Tarani Chandola et al., "The Effect of Short Sleep Duration on Coronary Heart Disease Risk Is Greatest Among Those with Sleep Disturbance: A Prospective Study from the Whitehall II Cohort," *Sleep* 33, no. 6 (2010): 739–44, https://doi.org/10.1093/sleep/33.6.739.

101. Yan Lu et al., "Association Between Sleep Duration and Cancer Risk: A Meta-analysis of Prospective Cohort Studies," *PLOS ONE* 8, no. 9 (2013): e74723, https://doi.org/10.1371/journal.pone.0074723.

102. Sandra Sephton and David Spiegel, "Circadian Disruption in Cancer: A Neuroendocrine-Immune Pathway from Stress to Disease?" *Brain, Behavior, and Immunity* 17, no. 5 (2003): 321–28, https://doi.org/10.1016/s0889-1591(03)00078-3.

103. Scott Davis, Dana K. Mirick, and Richard G. Stevens, "Night Shift Work, Light at Night, and Risk of Breast Cancer," *Journal of the National Cancer Institute* 93, no. 20 (2001): 1557–62, https://doi.org/10.1093/jnci/93.20.1557; Eva S. Schernhammer et al., "Rotating Night Shifts and Risk of Breast Cancer in Women Participating in the Nurses' Health Study," *Journal of the National Cancer Institute* 93, no. 20 (2001): 1563–68, https://doi.org/10.1093/jnci/93.20.1563.

104. Marie-Élise Parent et al., "Night Work and the Risk of Cancer Among Men," *American Journal of Epidemiology* 176, no. 9 (2012): 751–59, https://doi.org/10.1093/aje/kws318.

105. Méyomo Gaelle et al., "Night Work and Prostate Cancer Risk: Results from the EPICAP Study," *Occupational and Environmental Medicine* 75, no. 8 (2018): 573–81; Dapang Rao et al., "Does Night-Shift Work Increase the Risk of Prostate Cancer? A Systematic Review and Meta-Analysis," *OncoTargets and Therapy* (2015): 2817–26, https://doi.org/10.2147/OTT.S89769.

106. Elizabeth M. Ward et al., "Carcinogenicity of Night Shift Work," *The Lancet Oncology* 20, no. 8 (2019): 1058–59, https://doi.org/10.1016/S1470-2045(19)30455-3.

107. Norman B. Anderson et al., *Are Teens Adopting Adults' Stress Habits?* (American Psychological Association, 2014), https://www.apa.org/news/press/releases/stress/2013/stress-report.pdf.

108. Meena Kumari et al., "Self-Reported Sleep Duration and Sleep Disturbance Are Independently Associated with Cortisol Secretion in the Whitehall II Study," *The Journal of Clinical Endocrinology & Metabolism* 94, no. 12 (2009): 4801–9, https://doi.org/10.1210/jc.2009-0555.

109. Carli et al., "A Newly Identified Group of Adolescents."

110. Rebecca A. Bernert et al., "Objectively Assessed Sleep Variability as an Acute Warning Sign of Suicidal Ideation in a Longitudinal Evaluation of Young Adults at High Suicide Risk," *The Journal of Clinical Psychiatry* 78, no. 6 (2017): e678–87, https://doi.org/10.4088/JCP.16m11193.

111. Pierre A. Geoffroy et al., "Sleep Complaints Are Associated with Increased Suicide Risk Independently of Psychiatric Disorders: Results from a National 3-Year Prospective Study," *Molecular Psychiatry* 26, no. 6 (2021): 2126–36, https://doi.org/10.1038/s41380-020-0735-3.

112. T. Pereira, Sónia Martins, and Lia Fernandes, "Sleep Duration and Suicidal Behavior: A Systematic Review," *European Psychiatry* 41, no. S1 (2017): s854, https://doi.org/10.1016/j.eurpsy.2017.01.1699.

3
NUTRITIONAL WELLNESS

Introduction

Over the course of this chapter, you will learn about what can be changed regarding nutritional behavior by using the Change Assessment Model from chapter 1. This chapter includes three parts. *Part 1: Nature and Nurture* is about what you cannot change (nature) and what you can change (nurture). *Part 2: Deficiency and Excess* identifies the choices you have about the things you can change. *Part 3: Micro and Macro* is about the relationship between smaller and larger systems that you interact with. Those relationships include the cell and the body as well as the individual and the group. Your reality can be shaped by how you navigate these three areas.

Part 1: Nature and Nurture

What You Cannot Change (Nature)

There are certain nutritional principles that cannot be changed because they are part of human biology. Three of these principles are nutrient requirements, nutrient function, and nutrient impact. Although you are an individual and unique in certain ways, your body is affected, for better or worse, by these three principles. Once you understand these principles, you will be better able to appreciate the nutritional choices you can make.

Nutrient requirements
You cannot change the fact that the human body requires certain materials in order to sustain itself and operate the systems that allow the body to function. Those materials come in two basic categories: caloric or noncaloric. Caloric

nutrients have an energy value and can be used as fuel. Noncaloric nutrients have no energy value and cannot be used as fuel no matter how much you consume. Both types of nutrients are required for the body to survive and grow. Table 3–1 separates the caloric from the noncaloric nutrients.

Your body obtains calories when you consume carbohydrates, fat, protein, or alcohol. These caloric nutrients will fuel your body, but they all work somewhat differently. Alcohol is listed as a caloric nutrient, but your body does not require it the way that it requires carbohydrates, fat, and protein. That being said, alcohol in moderation has been repeatedly shown to be good for your health.[1]

Your body does not obtain calories when you consume vitamins, minerals, antioxidants, phytochemicals, or water. However, all of these nutrients are essential for a healthy life. Having noncaloric nutrient deficiencies or inadequacies can lead to severe illnesses that impact your risk factors for the Big Three (heart attack, cancer, and suicide) by affecting the following parts of your body:

- Heart[2]
- Cells (cancer)[3]
- Brain[4]

TABLE 3–1. Caloric and noncaloric nutrients

Caloric nutrients		Noncaloric nutrients
Carbohydrate	4 calories per gram	Vitamins
Fat	9 calories per gram	Minerals
Protein	4 calories per gram	Antioxidants and phytochemicals
Alcohol	7 calories per gram	Water

Nutrient function

You cannot change the fact that the caloric and noncaloric nutrients have specific functions. Caloric nutrients have different jobs in your body than noncaloric nutrients. The way these nutrients function is consistent, which is why nutrition is a science. Table 3–2 lists the caloric nutrients and provides a brief explanation of how they function.

The main function of caloric nutrients is to give your body fuel and to provide raw material for your body to rebuild itself. Your body can store excess fuel whether it is from fat or sugar in your diet, but it cannot store excess protein or alcohol. A limited supply of fat can be stored as triglycerides in your muscles, and any excess triglycerides are stored as body fat.[5] A limited supply of sugar can be stored as glycogen in your muscles and liver, and any excess sugar is also stored as body fat.[6] Some of the excess protein you consume can be converted into sugar or fat to be used or stored.

TABLE 3-2. Caloric nutrients and their function

	Caloric nutrients
Carbohydrate	Used immediately or stored as glycogen (sugar) in your muscles if possible, otherwise stored as body fat.[7] Fiber does not have a lot of calories, but fiber from plants can help to clean blood vessels.[8] Fiber can clean the gut.[9] It can also foster healthy gut bacteria.[10]
Fat	Used immediately or stored as fat in your muscles if possible, otherwise stored as body fat.[11] Fat from diet is used as cholesterol to rebuild almost all cells.[12] Fat is also used to make hormones.[13]
Protein	Used immediately as a building block to rebuild almost all cells.[14] Used to make enzymes and antibodies.[15] It also makes hormones.[16] Extra protein can be converted into glucose (sugar) or fatty acids (fat) and stored as fat if needed.[17]
Alcohol	Used immediately as fuel and prioritized over other caloric nutrients for metabolism.[18] Alcohol cannot be stored. In moderation, alcohol can help prevent heart disease.[19] It can help prevent dementia.[20] It can even help prevent some cancers.[21]

Your body can learn to store more triglycerides or glycogen in your muscles based on your fitness habits.[22] As you increase your endurance exercises, your muscles can adapt to store more triglycerides. As you increase your exercise intensity, your muscles can adapt to store more glycogen. The amount of fat or sugar that your body can store in your muscles for use during exercise depends on the type and the frequency of conditioning that you engage in. Ultimately, you will gain body fat if you eat more calories than you burn, and lose body fat if you eat less calories than you burn. As a reminder, genetic and environmental factors also play a role in the amount of calories you burn or store as body fat.[23]

Table 3-3 lists the noncaloric nutrients and briefly describes their functions. The main function of noncaloric nutrients is to support the regulation of your growth, repair, and system functionality. Noncaloric nutrients support your immune system and help to fight cancer. Noncaloric nutrients also support hormone production, enzyme production, and cellular function.

Nutrient impact

You cannot change the fact that nutrients have an impact on your body systems. In the book *Eat to Beat Disease*, author Dr. William Li explains how your nutritional behavior has an impact on five major systems that play a role in your health: the microbiome, DNA, stem cell system, immune system, and angiogenesis. Table 3-4 lists those five systems and provides a brief explanation of their functions.

TABLE 3-3. Noncaloric nutrients and their function

	Noncaloric nutrients
Vitamins	Must be consumed by diet, with the exception of three: K, biotin, and D.[24] Vitamins come in different amounts from different foods, so a diverse diet is necessary. Vitamins impact metabolism, organ functionality, DNA production and repair, red and white blood cell formation, and neurological functionality, just to name a few.[25]
Minerals	Must be consumed by diet. Minerals come from the earth and are transferred by plants and water to animals and humans. Minerals impact the health of your bones, heart, muscles, and skin, as well as nerve function, metabolism, acidity levels, DNA production, and sex hormone production, just to name a few.[26]
Antioxidants and phytochemicals	Found in natural foods and can help to prevent the oxidation of cells.[27] They also support different ways that the body fights cancer, such as apoptosis, angiogenesis, and autophagy.[28] As diet diversity increases, oxidation-clearing and cancer-fighting substances increase.
Water	Essential for brain function, muscle function, organ function, and cell function.[29] Critical for waste removal, temperature regulation, and blood volume.[30] The rule of thumb is to consume half of your body weight in ounces of water each day, not including water intake from exercise.[31]

TABLE 3-4. Five major body systems impacted by nutrition

	Five major body systems and their functions
Microbiome (gut)	The living system of bacteria in your intestines that help you break down your food.[32] Your microbiome can synthesize vitamins.[33] It can support your immune system.[34] It can even support your mental health by releasing chemicals like oxytocin, serotonin, gamma-aminobutyric acid (GABA), and dopamine.[35] The bacteria in your gut eat what you eat. The more diverse the food, the better your microbiome becomes.[36]
DNA	The code that tells each of your body's cells how to develop and function. When DNA is damaged, errors can occur in the way your cells develop and function. The dietary choices you make can impact how your DNA repairs itself.[37] Your choices can also activate or deactivate genes through epigenetics.[38] You can even repair or grow your telomeres at the ends of your DNA strands.[39]
Stem cells	Cells that can become almost any cell in order to help your body repair and rebuild.[40] Cells in the small intestine regenerate every 2–4 days, immune cells regenerate every 7 days, and cells in the lungs and stomach regenerate every 8 days.[41] If your body stopped producing stem cells, you would likely die in about a week.[42]

Five major body systems and their functions (continued)	
Immune system	A network of different types of cells that act to protect your body against both foreign invaders and internal issues like cancer. The immune system has an immediate but more generalized response (innate immunity) and a delayed but more surgical response (adaptive immunity).[43] Your diet can help to boost immunity and also reduce symptoms of autoimmunity.[44]
Angiogenesis	The process of forming new blood vessels. The body tries to maintain a balance of growing new blood vessels when needed and trimming blood vessels when not needed. When angiogenesis is either deficient or excessive, major issues can result such as cancer, heart disease, and obesity.[45]

What You Can Change (Nurture)

There are three factors of nutrition that you can control. You can control your nutrient variety, nutrient density, and nutrient quantity. The choices you make in these three areas will affect your health. This section will give you an overview of what these three factors are and why they should matter to you.

Nutrient variety
The first area of nutritional behavior that you can control is nutrient variety. Nutrient variety relates to the different types of caloric and noncaloric nutrients, the different food groups, and even the different types of food within each food group. You need to ensure that you are including all the different types of nutrients in your diet. Nutrient variety is critical because different nutrients serve different purposes in your body. Lacking the nutrients that you need will make it harder for your body to recover and for your many body systems to function normally. Nutrient variety includes an understanding about how to best prepare a plate for your meals in order to provide yourself with the best chance at obtaining all of the nutrients you need. The food pyramid was replaced with a much more intuitive guide called MyPlate which was developed by the U.S. Department of Agriculture (USDA) to help design well-balanced meals.[46]

Nutrient density
The second area of nutritional behavior that you can control is nutrient density. Nutrient density is a concept that Dr. Joel Fuhrman made famous in his book *Eat to Live*. It focuses on the ratio of nutrients to the number of calories in your food. For example, a 12-ounce sugary beverage might contain 150 calories, all from high-fructose corn syrup, and provide no vitamins, minerals, or other

health-promoting nutrients. A can of solid white tuna in oil has about the same calories as the soda but also contains fat, protein, and some vitamins and minerals. Nutrient density is about getting the most "bang for your buck" with each bite. Almost universally, whole foods from nature are more nutrient dense than processed and packaged alternatives. It is possible to fill yourself up with calories while starving your body of nutrients. To take it a step further, you can plan your meals with nutrients that can help you prevent and fight cancer. Dr. William Li developed www.eattobeat.org, which includes an online library of cancer-fighting foods.

Nutrient quantity
The third area of nutritional behavior that you can control is nutrient quantity. Nutrient quantity is about how much you need to eat based on your body size and activity level. Part of the explanation for how much you eat comes down to habits. You may have been taught that you should have three, five, or even seven meals a day. Knowing how to determine how much food you actually need is important if you are trying to lose, gain, or maintain weight. Additionally, having a strategy for accounting for your progress is an important part of managing your weight. But nutrient quantity does not just apply to calories. It also applies to water, as well as noncaloric nutrients. Firefighters should be aware of the potential dangers of extreme amounts of noncaloric nutrients that can come from supplements, which will be examined in part 2 of this chapter.

Part 2: Deficiency and Excess

Nutrient Variety
The first choice regarding your nutritional behavior is nutrient variety. Your meals should be the primary source of the different types of caloric and noncaloric nutrients that you eat. As a firefighter, you are not expected to be a nutritionist or chef, but you should know how to make a healthy plate of food for yourself and your shift. Making a healthy plate requires you to know the basic food groups and the recommended servings for each food group. This section will provide guidance on navigating the different caloric and noncaloric nutrients, the different food groups according to the U.S. Department of Agriculture's (USDA) MyPlate model, and the different types of foods within those food groups.

Caloric and noncaloric nutrients
There are many different types of diets that promise all kinds of results. When it comes to the four caloric nutrients (carbohydrate, fat, protein, and alcohol), the phrase "everything in moderation" serves as a good guide. The Mediterranean diet, which is not really a diet but more of a nutritional lifestyle, has been rated one of the healthiest "diets" for decades.[47] The Mediterranean diet consists of a balance of all the different caloric nutrients, including alcohol, which for most people will provide all the noncaloric nutrients you will need.

It is important to know that carbohydrates, fat, and protein often exist together in food. For example, 3 oz. of tenderloin steak has 26 grams of protein but also 7.6 grams of fat (and fat is not a bad thing). One cup of oats has 55 grams of carbohydrates, 11 grams of protein, and 5 grams of fat. One cup of whole milk has 8 grams of protein, 8 grams of fat, and 12 grams of carbohydrates. Choosing to exclude one of the caloric nutrients results in limiting the different building blocks that your body needs to grow and sustain its health. Additionally, you would end up excluding many different types of food, since many foods have a combination of all three caloric nutrients. Furthermore, excluding one of the caloric nutrients from your diet will make it harder for you to obtain the many different types of vitamins, minerals, and phytochemicals that your body needs.

It should be noted that while alcohol is a nutrient group, avoiding alcohol will not diminish your ability to obtain all of the caloric or noncaloric nutrients that you need to be healthy. Some people (East Asian people in particular) have a natural intolerance to alcohol and break out in a rash when they consume it.[48] If you have this alcohol-flushing response, you should know that there is a connection between this reaction and a heightened risk for esophageal cancer from alcohol consumption. Others may have sworn off of alcohol for health or lifestyle reasons. For those that do drink alcohol, a moderate amount has been shown to reduce risk of all-cause mortality by 20% and reduce risk of cardiovascular disease (CVD) mortality by 25%.[49] However, heavy drinkers and binge drinkers are at an increased risk of all-cause mortality and cancer.

Published guidelines can provide you with recommendations on how much of each nutrient you should consume as an adult. The Institute of Medicine suggests that you should consume the following:[50]

1. 45%–65% of your calories as carbohydrates
2. 20%–35% of your calories as fat
3. 10%–35% of your calories as protein

Depending on your lifestyle, you may need to adjust your intake of each of those groups within the ranges provided. It is important to realize that carbohydrates include fruit, vegetables, and grains, which is why it makes up such a large proportion of the recommended daily intake. Consuming less than the minimum recommendation would be considered deficient, and consuming more than the recommended maximum would be considered excessive. If managing these numbers in your head each day seems like a daunting task, relax. Looking at food groups instead of nutrient groups is a much easier way to manage your daily intake of food.

Food groups

The USDA introduced the MyPlate model in 2011. This model was a huge improvement from the food pyramid, a model which often left people confused about how to apply the different food group recommendations to their daily lives. MyPlate simplifies the idea of nutrient variety by using an easy-to-understand plate model to guide you as you craft each of your meals (fig. 3–1).

In the MyPlate model, you see the five major food groups represented on the plate: fruits, vegetables, grains, protein, and dairy. In general, the advice is to aim for the following proportions:

- ½ of your plate as fruit or vegetables
- ¼ of your plate as protein
- ¼ of your plate as grains

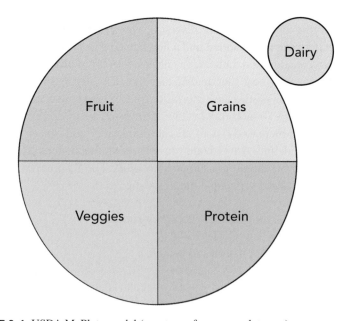

FIGURE 3–1. USDA MyPlate model (courtesy of www.myplate.gov)

A serving of dairy is depicted as additional to the plate. However, there are also healthy oils, fats, herbs and spices, and even sweeteners that do not make up major parts of a meal but are still important. The MyPlate website focuses on the five major food groups and provides helpful resources for shopping and cooking. Later in this section, you will be provided with additional resources.

The MyPlate model is a best practice, but it does not mean that every single meal you eat has to follow the MyPlate model. For example, you might consume fruit but not vegetables for breakfast. Conversely, you might have vegetables but no fruit for dinner. Sometimes you might be out of fruits or vegetables for a few days until you go shopping again. Occasionally, you may want to treat yourself with a pizza night. However, your choice to ignore a specific food group would mean that you are deficient in that food group, and your choice to overemphasize a specific food group would mean you are excessive in that food group. This can be detrimental to your health in the long term.

MyPlate should be interpreted as a goal that your overall nutrition patterns match fairly closely. The important take-home message is to aim for the appropriate ratio of foods from the different groups, not consuming too few or too many sources from each area of MyPlate. Any food group within the MyPlate model can be consumed either deficiently or excessively. When the food groups are either deficient or excessive, your nutrients can get out of balance.

Some people avoid certain food groups for perfectly acceptable reasons. For instance, you might have a sensitivity or intolerance to dairy and so you never consume dairy. It may be against your religion or lifestyle to eat meat and so you have to find other ways to get the right amounts of amino acids from proteins. You will be considered deficient in any food group if you consume less than the recommended amount or consume none at all. However, that does not mean your health has to suffer. It just means that you should be aware that you may be missing certain nutrients in your diet and will need to ensure that you are obtaining those nutrients elsewhere.

The different food groups represent the basic variations of food, but variety does not stop there. If you eat the same exact type of food from each food group, not only would it be boring after a while, but you would be limiting the types of nutrients that you are eating. Therefore, you should consider the different types of foods that exist within each food group, as well as the different colors of foods that exist, and make an effort at increasing those varieties as well.

Types and colors of food

Within the five major food groups, there are many different types of food. Different types of food within each food group provide different types of caloric and noncaloric nutrients. Variety in food is fun, but as an industrial athlete, firefighters should look at food as fuel and not just as fun. Putting aside food that is against your religion, lifestyle, or ability to digest, you should be interested in increasing your food variety to boost your health. Having three to five types of food for each food group is a great way to have variety. Table 3–5 provides examples of different types of foods within each of the five major food categories.

TABLE 3–5. Examples of different types of food within each food group

Group	Fruits			Vegetables		Proteins		Grains		Dairy	
Food	Banana	Blueberries	Sweet potato	Broccoli	Salmon	Beef	Rice	Oats	Milk	Yogurt
	Strawberries	Apple	Kale	Carrots	Chicken	Pork	Wheat	Quinoa	Cheese	Butter/cream

Color is another way that foods are differentiated. The different colors of food usually represent different caloric and noncaloric nutrients. The color of food will obviously change between different types of food, but it can also change within a specific type of food. For example, fish and poultry are different types of food and have different colors. However, there are different colors of fish and there are different colors of poultry (light meat and dark meat). The colors of food are a major indicator of nutrient content.

As a general rule of thumb, the darker the color, the more nutrients that are available in that food. However, the key is to choose a variety of colors, because each color provides different nutrients in your diet.[51] Table 3–6 provides an example of different noncaloric nutrients for different food colors.

TABLE 3–6. Different nutrients from different colors of food—adapted from Katherine D. McManus in Harvard Health Publishing[52]

Color	Noncaloric nutrient	Benefit
Red	Lycopene	Protects against prostate cancer and heart and lung disease
Orange and yellow	Beta-cryptothanxin	Supports intracellular communication and may also help prevent heart disease
Green	Sulforaphane, isocyanate, and indoles	Inhibit carcinogens (cancer causing agents) in the body
Blue and purple	Anthocyanins	Delay cellular aging and prevent blood clots
White and brown	Allicin, quercetin, kaempferol	Fight cancer and other chronic diseases

Dr. Li explains that the more variety or diversity in your diet, the healthier your microbiome will be; this is because the bacteria in your gut eat what you eat, and different foods allow different bacteria to survive and grow. You should care about the health of your gut bacteria because your microbiome plays a major role in producing chemicals that affect your mood.[53] Your gut bacteria also help to synthesize vitamins.[54] Your gut even supports your immune system.[55]

It should be noted that taking antibiotics not only kills bad bacteria in your body, but it also kills good bacteria, especially in your gut. Therefore, if you take antibiotics for an illness, you should put extra effort into eating healthy to restore your good gut bacteria. Failure to restore or maintain good gut bacteria can cause harmful bacteria to take over your gut, which can cause many types of health issues.[56]

Deficiency and excess regarding nutrient variety manifests in a few ways. The first manifestation involves caloric and noncaloric nutrients. Ignoring any one of the caloric or noncaloric nutrient groups is deficient, and focusing too much on any of the groups is excessive. As an example, swearing off of all carbohydrates may work temporarily to help you lose weight, but it also means removing fruits, vegetables, and whole grains out of your diet, which provide many of the noncaloric nutrients that your body needs to stay healthy. On the other hand, eating only carbohydrates (fruits, vegetables, grains) will make it harder for you to get the necessary fats and proteins that your body needs to function. The key is to have a balance of the different caloric nutrients.

The second manifestation involves the basic food groups. Ignoring any one of the food groups is deficient, and focusing too much on any of the food groups is excessive. As an example, even if you eat carbohydrates as a nutrient group, if you exclude fruit, you are missing out on many nutrients, antioxidants, and phytochemicals that fruits offer. However, if fruits are your only source of carbohydrates, you will miss out on the many noncaloric nutrients that come from vegetables and whole grains.

The third manifestation involves food types and colors. Each person has their own preferences, and when it comes to food types, there are plenty of varieties to choose from. You can have nutrient variety without having to eat every type of food that exists. As an example, if you eat carbohydrates and you eat fruits, you might eat apples, bananas, and some berries. You are not expected to eat all of the different fruits all the time. However, you should make an effort to have different colors of food as part of your nutrient variety. Excluding one or more colors of food would be deficient, and eating too much of one color would be excessive. You do not have to eat all colors for all food groups, but the more colors you eat, the better.

Nutrient Density

The second choice regarding your nutritional behavior is nutrient density. The caloric nutrients in your food come from carbohydrates, fat, protein, or alcohol, and noncaloric nutrients in your food come from vitamins, minerals, and phytochemicals. Dr. Joel Fuhrman explains in his book *Eat to Live* that nutrient density is the total amount of nutrients divided by the amount of calories (Nutrient Density = Nutrients/Calories).[57] Your choice in foods will determine whether you are getting a high proportion of nutrients per calorie or whether you are mostly just getting calories without nutrients. The idea of nutrient density is especially relevant in today's society, where processed food is so easy to get and it feels like you have no time to buy and prepare natural food. This section will provide guidance on navigating nutrient density regarding how foods are processed, how foods are raised, and the science behind assigning different values to food based on their nutrient content.

Understanding how food is processed

The idea of processed food means a few different things. Processed can refer to any food that has been changed in some way. Some examples include the following processes:

- Shredded carrots have gone through a shredding process and can be found in a bag
- Frozen spinach has been cut and flash frozen and can be found in a bag
- Corn in a can has been taken off the cobb and placed in a tin container in salt water
- Bacon is sliced pork belly that is seasoned and often preserved with chemicals
- Vegetable oils are often processed using chemical extraction methods

Usually when someone talks about processed food, they are not referring to natural fresh produce that has been shredded, or even natural food that has been frozen. The idea of processed food usually refers to food that either no longer looks like natural food or has added chemicals. However, it is important to understand that processing exists on a spectrum, from minimal (e.g., shredding carrots) to fully processed (e.g., candy bars). And you need to know the differences.

Processed food means that what you are eating has been changed in some way so that the food is not exactly the way it was found in nature (fig. 3–2). Some of the different ways food is processed include the following:

- Mechanical: shredded, chopped, diced, ground, pressed
- Chemical: flavors, colors, preservatives, emulsification, hydrogenation
- State: frozen, dried, canned, pickled, fermented, raw, cooked

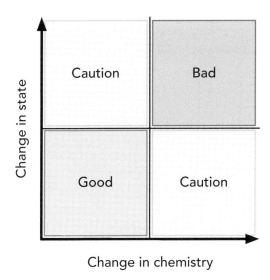

FIGURE 3-2. Processed food matrix

Not all ways of processing food are unhealthy. As a general rule of thumb, it is best to buy food the way it comes from nature, such as fresh produce or freshly cut meat from the butcher. However, some foods that are processed make our lives easier without making our meals less healthy. Having food that is precut and bagged, or even flash frozen, allows us to save time without changing the nutritional value of the natural food. Also, some foods that have been fermented are actually healthy for you, such as yogurt and kimchi, because they support your gut bacteria. In general, the farther your food gets from nature (fresh or frozen), the more likely it is low in nutrient density.

When most people talk about processed foods, they are referring to manufactured food products that can last for a long time in your pantry. These types of processed foods are normally found in the aisles of the grocery store in a box, bag, can, or jar and can stay on the shelves for a longer period of time than natural food because of added preservatives. The processed food usually includes added sugars, starches, and fats in order to make the food tastier or to help keep the food product together in a specific shape.

Table 3-7 lists examples of processed foods compared to a natural alternative. When using 100 grams as the measurement of comparison, the processed food has much higher calories than what you would get from nature alone. Even if the amount of nutrients was the same in natural and processed foods, the processed foods have more calories and therefore a lower nutrient density. In addition, it is likely that the extra calories from processed foods are coming from added sugars.

TABLE 3-7. Energy density comparison of different foods*

Group	Fruits		Vegetables		Proteins		Grains		Dairy	
Food	Dried apple	Apple	Tomato soup	Tomato	Chicken nugget	Chicken thigh	Rice cracker	Brown rice	Milk-shake	Milk
Calories per 100 grams	243	52	72	35	296	177	384	111	112	61

*100 grams could represent more or less than a typical serving of these foods. We used 100 grams as a way to standardize the weight of each food.

One critical nutritional behavior you can do to increase your nutrient density is to read nutrition labels. Nutrition labels do not exist on fresh produce. Sometimes your butcher will provide a generic nutrition label for fresh meats and fish, but there is only one ingredient—the meat. When you do decide to eat processed food as a treat or you need to grab something in a hurry, you should be reading the ingredients. If you pay attention, you will notice that processed foods usually have the first ingredient as some sort of sugar. As a general rule of thumb, the more ingredients you see listed on the food, the more processed it is. Another rule of thumb is that if you cannot understand what the ingredient is, it is likely a chemical that is not good for you.

To make things more complicated for you as a consumer, food companies use many different types of sugar and other processed ingredients on the nutrition labels that can confuse you. Some of the words used instead of "sugar" include dextrin, dextrose, fructose, glucose, high-fructose corn syrup, maltodextrin, maltose, and sorghum, just to name a few. It is also worth noting that in many processed foods, companies will "enrich" them by adding vitamins and minerals because the food would have hardly any of those nutrients if they were not added. You should be cautious of the food when you see these types of additives because it means the food is highly processed.

There are no national guidelines about how much processed food compared to how much natural food you should eat. Ideally, all the food you eat would be natural. It is a given that humans enjoy entertainment and going out to eat, and they sometimes have to get food on the go. Therefore, there is no standard way to determine what constitutes a deficiency of natural foods and what constitutes an excessive amount of processed foods. However, a positive correlation exists between the amount of processed food you consume and your risk of heart disease.[58] The same is also true for cancer.[59]

Eating natural food is almost always going to be better than eating processed food. However, the story does not end there. It is important to realize that those who produce food have an incentive to increase production in order to increase profits. For that reason, you should inform yourself about the different ways food is raised.

Understanding how food is raised

All food is not raised in the same way. Farmers use different strategies to raise the animals or crops they sell. You should be aware that food is a business, and some farmers do not care about the quality of the food you eat and instead prioritize the quantity of food they can sell. You have probably heard the saying, "You are what you eat." However, that idea also applies to animals and plants. For that reason, another saying exists, "You are what your food eats." Therefore, even natural, unprocessed food can have different levels of nutrients depending on how it was raised:

- The soil quality will impact the quality of fruit, vegetables, and grains
- The feed quality will impact the quality of meat, fish, and eggs
- Fruit, vegetables, and grains may be heavily sprayed with chemicals
- Meat and fish may have been raised in a habitat that is inhumane or unnatural
- Meat and fish may have been injected or fed antibiotics, hormones, and other chemicals
- The fruit, vegetables, and grains may be genetically modified organisms (GMOs)
- Meat and fish may be GMOs or may have been fed GMO products

Another critical nutritional behavior you can do is to consider the quality of the food and how it was farmed or raised. There are many different types of branding that are used to signal to the customer that a food is "more" natural. Phrases like "All Natural," "Non-GMO," and "Certified Humane" are just some of the examples of branding that are used to sell products. "Organic" is another type of labeling and is regulated by the USDA. It is important for you to understand that all of these labels mean different things. Table 3–8 provides a brief explanation of each label.

TABLE 3-8. Navigating the meaning of different food labels

Label	Meaning (U.S. food regulations)
All natural	The FDA has no formal definition for the term "natural." The FDA considers the term "natural" to mean nothing artificial or synthetic has been added (including food coloring). There is no standard on the use of pesticides or food production methods when using the label.[60] This label does not mean the food is raised well.
Non-GMO	The Non-GMO Project is a third-party organization that has published a standard for food products. The non-GMO label does not mean that the food is 100% GMO free. It also does not mean the product is safer, better, or healthier. It simply means the product complies with the Non-GMO Project Standard.[61]
Gluten-free	The FDA regulates gluten-free labeling. The label means that the food product cannot be intentionally made with any amount of gluten-containing grain or an ingredient derived from the grain without having first removed the gluten.[62] The label does not mean the food is healthy for you.
Free range	The FDA regulates the free range label for use on poultry only. The label simply means that birds are given access to outdoors, but the time spent outdoors is not regulated. The free range label is not regulated at all for use on eggs.[63]
Cage free	Cage free is a label used on egg cartons and means that the laying hens were not kept in cages. It does not mean the hens were allowed to go outdoors. The FDA requires that food labels be truthful but has no definition of cage free.[64]
Pasture raised	Pasture raised is a label used on meat, poultry, dairy, or eggs and means that the animals were given access to pasture for at least some portion of their lives and not continually confined indoors. However, this label is not a government-regulated label and there are no standard definitions for what this means.[65]
Certified humane	Certified humane is a label that comes from a 501(c)(3) nonprofit. The label indicates that the producer meets the animal care standards from birth to slaughter. This includes never keeping animals in cages, crates, or tie stalls and allowing the animals to do what comes naturally. Animals must be fed quality feed without antibiotics or hormones.[66] Certified humane animals can be fed GMO products.
Organic	The USDA regulates organic labeling, which means that animals were raised without antibiotics or hormones and cannot be fed GMO products. Plants are produced without conventional pesticides, synthetic fertilizers, sewage sludge, bioengineering, or ionizing radiation. In order to qualify for the USDA organic seal on a food product, the product must have at least 95% organic ingredients.[67]

The way food is raised can have a major impact on your health. One high-profile and far-reaching example is the use of the herbicide Roundup, which was manufactured by Monsanto. Glyphosate is the main ingredient in Roundup. In 2015, the International Agency for Research on Cancer (IARC) classified glyphosate as a "probable carcinogen" to humans.[68] Many lawsuits against Monsanto (the original owner) and Bayer (the new owner) have followed, including an $11 billion settlement that covers roughly 100,000 claims.[69]

The U.S. Environmental Protection Agency (EPA) has stated that glyphosate is "unlikely" to be carcinogenic.[70] The new owner, Bayer, has stated they will stop using Roundup for residential use in 2023. However, Roundup will still be available for agricultural use. Roundup is historically the most popular herbicide in the U.S.[71]

When buying food from a grocery store, the USDA organic seal is the gold standard. Buying foods that are organic means you will greatly reduce your exposure to pesticides, herbicides, fungicides, antibiotics, hormones, GMOs, or other chemical additives that are not healthy for you. However, organic can be expensive and you may not be willing to pay the additional cost for everything you buy. Therefore, you should know that information exists on which nonorganic food items have the most chemical residue. That information is publicly available.

There is a nonprofit organization called Environmental Working Group (EWG) that publishes a list of fresh produce called the "Dirty Dozen." The EWG analysts look at USDA data for chemical pesticide residue levels. The USDA tests are done after washing, scrubbing, and peeling the produce as customers would. Still, nearly 70% of all nonorganic fresh produce has harmful chemical residues. The Dirty Dozen list is published annually to warn consumers about which nonorganic fresh foods have the highest levels of chemicals. The list can be found on EWG's website at www.EWG.org. In 2022, the list from 1 to 12 included: strawberries, spinach, kale/collard/mustard greens, nectarines, apples, grapes, bell and hot peppers, cherries, peaches, pears, celery, and tomatoes.

Hippocrates is attributed with saying "Let food be thy medicine, and let medicine be thy food." You must navigate through a lot in order to use food as "medicine." Even if you buy fresh food with only one ingredient, chemicals may be on it or may have been fed to it. However, your efforts at eating nutrient-dense foods are a key part in using food as medicine and avoiding health issues down the road. For that reason, it is important to understand what is in the food you eat.

Understanding nutrient content

Each food item has different caloric and noncaloric nutrients, and thus will affect your body in different ways. You may have no desire to remember what types of nutrients are in which foods, but that is not the goal here. Instead, there are databases that provide useful information on different types of food and the nutrients they contain. Being aware of some of these databases and using them to help you create a shopping list of nutrient dense foods is a simple way to optimize your nutritional behavior. Three databases that can help you choose foods are the University of Sydney Glycemic Index, Dr. Fuhrman's Aggregate Nutrient Density Index (ANDI), and Dr. Li's Universal Health Atlas[SM] cancer-fighting food database.

Glycemic Index (GI)
The GI is a standardized system (from 0–100) that rates foods on how much impact the food has on your blood sugar levels compared to pure glucose. It is not clear whether a high-GI diet impacts your body weight.[72] However, it is clear that a high-GI diet increases your risk of Type 2 diabetes.[73] It also increases your risk for cardiovascular disease (CVD).[74] Individuals who eat a high-GI diet and who also have a history of CVD increase their risk of all-cause mortality by 51%, but even individuals without a history of CVD who eat a high-GI diet increase their risk by 21%.[75]

Foods are separated into three categories—low, medium, or high GI. In general, the more processed the food is, the higher the GI. As an example, Wonder Bread has a GI of 71, which is higher than table sugar with a GI of 65. Low-GI foods are not necessarily more nutrient dense than high-GI foods. However, eating less processed food will bring your average GI level down per meal, and natural foods in general have a higher nutrient density than processed food. The University of Sydney (Australia) maintains a GI database. The database is available to anyone with internet access at www.glycemicindex.com.

Aggregate Nutrient Density Index (ANDI)
The ANDI, created by Dr. Joel Fuhrman, is a method of rating the nutrient density of food. Dr. Fuhrman argues that the nutrient density in your body comes from the nutrient density in your food, and therefore your diet is a major factor in whether you will suffer from major lifestyle diseases. While it is necessary to get enough calories in your diet, the key to health according to Dr. Fuhrman is to maximize the amount of nutrients you eat while obtaining the necessary amount of calories. The ANDI rating method uses 34 nutritional parameters for its scoring process, which range from the basic nutrition facts to more complex components like phytochemicals.[76] A quick internet search for "ANDI foods" or "ANDI chart" will provide you with resources that you can review.

Eat to Beat Cancer
The Eat to Beat Cancer campaign is an initiative to help fight the cancer epidemic by eating foods that have cancer-fighting properties. Dr. William Li is the CEO and Medical Director of the Angiogenesis Foundation. The Angiogenesis Foundation is a nonprofit that focuses on fighting cancer and other diseases by controlling angiogenesis, which is the process of creating new blood vessels.

The foundation's team has developed a database of natural cancer-fighting foods with the evidence to back it up. The database can be found at www.eattobeat.org/food.

Applying what you've learned about nutrient density
Firefighters are becoming more aware of cancer. It is likely that you know a firefighter who has been diagnosed with or died of cancer. Firefighters are not bashful about cursing cancer, but how much effort are firefighters, both as individuals and as a service, doing to prevent cancer through lifestyle changes? Fire companies that work 24 hours at a time normally have meals together. Does your shift drive to work in $70,000 pickup trucks and SUVs and then complain when the firefighter cooking dinner goes over budget because he bought some vegetables or a salad? Do you buy the cheap bacon with all the hormones, antibiotics, and nitrates (preservatives) because you refuse to spend the extra dollar on the good bacon? With the information in this section, you can create a shopping list of food that is scored low in the GI, high in nutrients per calorie, and high in cancer-fighting properties. Figure 3–3 is a shopping list that aligns with MyPlate and the food categories at www.eattobeat.org to help you easily create a list of foods that you know will help increase your health.

Deficiency and excess regarding nutrient density is manifested in a few ways. The first manifestation involves processing. Your goal should be to eat mostly whole, unprocessed food. It would be deficient to eat only processed food, but it would be excessive if you never allowed yourself to have any processed food. Processing involves many different types of alterations to food, and not all of them pose health risks. Cutting and freezing produce does not change the nutrient content. If you are trying to eat healthier, the first step is to reduce your processed food intake that involves added ingredients, preservatives, or chemicals.

The second manifestation involves how food is raised. The saying, "You get what you pay for," generally applies here. Selecting the food you eat based only on the price will usually result in getting food that is raised the cheapest way, which is often not the healthiest for you. Buying in bulk is also less expensive, but that is not the issue here. It would be deficient to never consider organic food—unless you are growing your own food or buying it from a trusted source—but it would be excessive if every food item you buy is organic. Keep in mind that food is a business with the goal to make money. If you are trying to eat healthier, the second step is to eat food that is raised or farmed well.

The third manifestation involves nutrient content. If you are trying to increase your nutrient density, then you need to be aware of which nutrients your food is providing for you. The level of processing and the level of care

Fruits	Veggies	Grains/seeds	Meats	Dairy	Nuts, legumes, beans	Oils/sauces	Spices and herbs

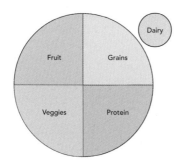

Drink	Sweeteners	Snacks

Cancer-fighting foods: www.eattobeat.org
My plate model: www.myplate.gov

FIGURE 3–3. Shopping list template

that went into raising your food will impact the nutrient density. It would be deficient to shop without any understanding of how to choose food based on nutrient content, but it would be excessive to never allow yourself to step out of those guidelines. High-GI foods, especially processed foods, generally offer lower nutrient density because those foods have a higher sugar content, resulting in higher calories. Foods with a high ANDI score are more nutrient dense, which means those foods offer more nutrients per calorie. The Eat to Beat Cancer database provides a list of evidence-based cancer-fighting foods. If you are trying to eat healthier, the third step is to be purposeful in how you shop for nutrient content.

Not everyone likes to research, even if it involves their own health. When it comes to creating a shopping list of high-quality natural foods, it is understandable that you may just want someone to tell you what to eat. However, the shopping list will likely be different for you as an individual than it is for your family or shift, and may require a discussion when a group is involved. In the firehouse, it is very easy to just go along with whatever the cook is preparing and withhold any complaints or comments. But having a discussion with the shift about choosing healthier foods might result in a surprising outcome. This is especially true if the food in question is the same, and you are simply swapping it out for a healthier version.

If you have no desire to look into the GI or ANDI scores, that is okay. At the very least, use the www.eattobeat.org database to making a shopping list. This way, you, your family, and your shift can feel satisfied that you are making an effort to prevent cancer.

Nutrient Quantity

The third choice regarding nutritional behavior is nutrient quantity. The reason this is the third choice is not because it is an unimportant concept. Eating too much food will cause you to gain weight, unless you are one of those genetically lucky individuals who never gains weight no matter how much you eat. The issue is that most people do not count their calories and have no desire to do so. Most people will simply adjust their food intake based on their hunger levels. Regardless, you should understand that a primary factor in weight gain comes down to how many calories you are eating compared to how many calories you are burning. This section will explore calculations on how much you should eat based on your height, weight, and activity levels. It will also explore hydration, supplement use, and an easy way to track weight loss.

Creatures of habit

There is no incentive to make things more complicated than they need to be. Having a routine helps to automate life. Routines allow you to spend less time and energy thinking about details of your day. This is a good thing because you can use that time and energy to think about your family, career, or projects that you need to accomplish that are not routine. However, when it comes to eating, your routine can cause the weight gain that you find so frustrating. Depending on where you are in your career, you may be holding on to a nutritional routine that fit a younger or more active version of yourself (fig. 3–4).

As you grew from an infant to a fully mature adult, your caloric needs increased as your body grew and activity levels increased. There is a time period after reaching adulthood that most people consider their "peak." Sometimes it feels like you can eat and drink anything you want during that time without any negative repercussions like weight gain. Then, your career starts, and more time is spent sitting, and less time is spent playing sports, hiking through the woods, or walking around the city. But by then you have become so used to your

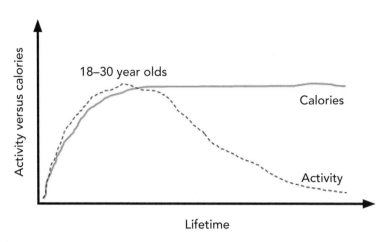

FIGURE 3-4. Imbalance between activity and calorie intake leads to weight gain.

nutritional habits that you continue to eat the same way, even as you do less and your body is no longer growing. This diversion is the beginning of weight gain for many people, even if they eat a relatively healthy diet. The weight gain is much more dramatic for those who eat mostly processed food. The latest data from the Centers for Disease Control and Prevention (CDC) shows that the U.S. obesity rate is 42% for the adult population.[77]

There is nothing wrong with gaining a few pounds as you age. A positive correlation exists, however, between the amount of extra weight you are carrying and your chances of cancer.[78] Therefore, while nutrient variety and nutrient density are critical for ensuring that your body is getting the nutrients it needs to function and stay healthy, nutrient quantity is a factor that plays a major role in obesity, which in turn increases your chances for cancer and other chronic diseases.

Calculations for nutrient needs

So, how much energy do you need? It depends! How much energy you need actually varies daily and by your age, height, and weight. Your estimated energy requirement (EER) is the sum of your resting metabolic rate (RMR), the thermic effect of feeding (TEF), and physical activity (PA), which is any movement above and beyond resting. The equation for energy needs is therefore EER = RMR + TEF + PA.

Most of the energy you need is relatively constant and represented by your RMR, or the amount of energy needed to support your basic bodily functions. Table 3-9 contains an equation to help you estimate your RMR.

Now that you know how to calculate the RMR portion of the EER equation, you need to know how to calculate PA. TEF represents the amount of

TABLE 3-9. Mifflin-St. Jeor equation for estimating RMR

Male: (10 × weight) + (6.25 × height) − (5 × age) + 5
Female: (10 × weight) + (6.25 × height) − (5 × age) − 161

Note: Weight is in kilograms, height is in centimeters, and age is in years. To convert pounds to kilograms, divide pounds by 2.2. To convert inches to centimeters, multiply inches by 2.54.

energy used to break down and absorb the food and beverages a person has consumed. TEF typically represents 5%–10% of EER but is not generally included as a variable. PA includes all movement above and beyond sitting. Your RMR is multiplied by an activity factor that represents your typical PA level. To refresh your memory about PA levels, review the fitness chapter section on exercise intensity.

PA is the most variable portion of your EER. However, even though your energy needs technically change daily, the overall pattern of meeting your energy needs is more important than minding precise daily numbers. Physical activity factors are shown in Table 3–10.

TABLE 3-10. Physical Activity (PA) factors

Description	PA factor	Examples
Sedentary	1.5	Little to no moderate physical activity (PA)
Low active	1.6	Moderate PA <150 min. or vigorous PA <75 min per week
Active	1.8	Moderate PA ≥150 min. or vigorous PA ≥75 min per week
Very active	2.0	Moderate PA ≥300 min. or vigorous PA ≥150 min per week

Note: Moderate activity = 3.0–6.0 METs, which is activity beyond leisure and up to about a stiff walk. Vigorous activity = ≥6.0 METs, which is any activity that is jogging or higher.

Here is an example of a 40-year-old, 6-foot-tall, 200 lb., male firefighter. We will assume a PA factor of 1.6 because he does about 30 to 60 minutes of moderate activity most days of the week.

> Example: 200 lb. divided by 2.2 = 90.9 kg
>
> 6'0" is 72 inches, multiplied by 2.54 = 182.88cm
>
> Using the equation in Table 3–9:
>
> (10 × 90.9) + (6.25 × 182.88) − (5 × 40) + 5 = 1,857 calories per day
>
> In order to account for PA: 1,857 × 1.6 = 2,971 calories per day based on age, weight, height, and PA level

You can now estimate your approximate energy needs. Fill in the blanks and complete the math:

> Men: (10 × __ kg) + (6.25 × __ cm) − (5 × __ years) + 5 = __ calories
>
> Women: (10 × __ kg) + (6.25 × __ cm) − (5 × __ years) − 161 = _____ calories
>
> Calories __ × PA Level __ = __ your typical energy needs

You may be thinking to yourself that you will never use this calculation. Even if you were to use that calculation daily, you still may not remember it. It is presented here in case you want an official reference point. However, there is a way to get a rough estimate as to how many calories you need per day. A publication by Harvard Health Publishing says that for those who have at least 30 minutes of moderate activity per day, it is as simple as multiplying your current weight by 15.[79]

> Easy Caloric Need Estimate: Weight × 15
>
> Example: 200 × 15 = 3,000 calories

As you can see, that estimate is very close to the number that was calculated using the Mifflin-St. Jeor equation (2,971 calories). The accuracy of how many calories you need each day will decrease if you use the easier method. The more serious you become at losing or gaining the next pound or two, the more important it may be to you to understand how many calories you need.

It is important to know that if you are trying to lose weight, the recommended amount of safe and sustainable weight loss is 1–2 lb. per week. To lose 1 lb., you have to decrease your calories by about 3,500 per week. You can do that by burning the calories or avoiding the calories in your diet. Therefore, if you achieve a 500-calorie deficit each day, whether by exercise or changing your diet, you will lose roughly 1 lb. in 7 days.

Calculation for hydration

Water is critical to optimal health. It makes up about 70% of your body weight. Because of this, dehydration will negatively impact you in many different ways. Some examples include a loss in physical performance, a loss in mental performance and concentration, and a reduced ability to regulate temperature.[80] However, just like calories, hydration levels can be deficient or excessive. Drinking too much water for your weight and activity levels can lead to a situation called hyponatremia, which is caused by having your sodium levels drop too low.

Firefighters are said to be industrial athletes. However, not every firefighter is where they should be from a physical preparedness standpoint. This book will use the hydration calculation from the American Council on Exercise (ACE) Health Coach Manual.

> Hydration: 1–1.5 ounces of fluid per kilogram of body weight (1 lb.=2.2 kg)

Prior to any exercise, the recommendation for athletes is to have 1–1.5 ounces of fluid per kilogram of body weight.[81] In order to figure out how much water you should be consuming, then, you need to figure out your weight in kilograms. To do that, simply divide your weight by 2.2. Once you have that number, multiply it by 1 and 1.5. These numbers become the upper and lower limits of your range.

> Example: 200 lb. (200/2.2) = 90.9 kg
>
> 90.9 x 1 = 90.9
>
> 90.9 x 1.5 = 136

The example shows that a 200 lb. firefighter should consume between 90.9 and 136 fluid ounces per day. This example is based only on weight and not environmental conditions or activity levels. The ACE Health Coach Manual also states that fluid replacement should be a 1:1 ratio. In other words, if you lose 1 pound of body weight during exercise, you should drink a pound (16 ounces) of water, because it is assumed that the weight you lost was from water.

There is an easier way to remember how much water you should be drinking daily. Drink half of your body weight in ounces of water each day. The number will fall within the range from the previous example. Again, this number does not account for exercise or environmental conditions.

> Easy Hydration Estimate: ½ your body weight in ounces of water
>
> Example: 200 lb. (200/2) = 100 ounces of water

The example shows that a 200 lb. firefighter should consume 100 fluid ounces per day. The upper range of the first estimate is still substantially higher than the second (easier) estimate—36 ounces is a little over a liter (or quart) of water. If you buy a water bottle that has the capacity for a liter or quart, then you can carry roughly 32 ounces of water on you at a time. If you drink your water bottle at each meal, you will have consumed 96 ounces of water in the day, which falls into the range of the first calculation and gets very close to the second calculation. This would provide your base level of daily hydration. You should be sure to hydrate above your baseline if you exercise, if it is really hot outside, or if you run a call where you put in some good work.

A note on supplements

The nutrition supplement industry is estimated to be worth upwards of $40 billion annually.[82] The nutrition supplement industry is also unregulated by the Food and Drug Administration (FDA). This means that the health claims on supplement labels may not be true, and the ingredients may be misleading or inaccurate. Even if the ingredients are accurate, the quantities of ingredients may not be accurate.

Some people who are deficient in a particular nutrient or who have a disease that affects their nutrition status may need to take a supplement. These individuals should be consulting their personal physician or dietician. If you have not been diagnosed with a nutrient deficiency, you should aim to get your vitamins, minerals, antioxidants, and phytochemicals from whole, unprocessed foods, especially from fresh produce.

Supplements provide you something in a way that nature never intended. By design, supplements isolate a particular substance, which is not how nature normally packages nutrients. Additionally, you risk taking an excessive amount of a particular substance that you would never get from nature. Vitamins and minerals are not the only type of supplements. Protein powders, meal replacement bars, energy drinks, and performance enhancement products are supplements too. Anything in excess is harmful, but when it comes to supplements that directly change your hormone production, there may be side effects even if you don't use the supplement in excessive amounts. Consider using your money for high-quality food instead.

Nutrient quantity impacts both your weight and your health. At the end of the day, weight management is highly dependent on the amount of calories you consume compared to the amount of calories you expend. Having a caloric deficit will cause you to lose weight, and having an excess amount of calories will cause you to gain weight. But, you need caloric nutrients to survive, and this section provided both a complex and a simple way to calculate how many calories you need. It also provided a complex and simple way to calculate how much water you need. Remember that hydration can be deficient, resulting in decreased mental and physical performance, but it can also be excessive, resulting in imbalanced electrolytes which can lead to serious health consequences.

It is worth repeating that the supplement industry makes lots of money, and unless you are deficient in a certain nutrient and have consulted with a professional, taking supplements is at best a poor use of your money and at worst could introduce excessive levels of isolated nutrients to your body, which could have unintended consequences.

Weight Management

Managing your weight can be a frustrating process. Some people have tried documenting everything they eat all day long, and that can make you go crazy. Improving your nutritional behavior is about changing habits, one step at a time. The accountability chart in figure 3–5 can be used to document behavioral changes that impact your weight. The concept is to focus on just two high-impact behaviors at a time to help you to move in the right direction. A pound is roughly 3,500 calories.

In order to lose 3,500 calories in a week, you have to lose an average of 500 calories per day. There are two columns on the chart; each represents a different goal (Goal 1 and Goal 2). Most people who are trying to lose weight know that there are foods or drinks that they are consuming that they probably should avoid. Removing these items would represent a high-impact behavior. Some people who are trying to lose weight are completely sedentary. Increasing movement is another high-impact behavior. Therefore, one goal can address diet and the other, fitness.

> Example: Goal 1 = Stop drinking sugary drinks
>
> Goal 2 = 10 minutes of running each day

The purpose of the chart is to collect data on your behavior and the outcome it causes. The chart has a space for you to document your weight at the beginning and end of each week. You can also choose to use a different metric such as body mass index (BMI) or body fat percentage (BF). It might also be

Month:					
Day	Goal #1	Goal #2	Day	Goal #1	Goal #2
1			16		
2			17		
3			18		
4			19		
5			20		
6			21		
7			22		
8			23		
9			24		
10			25		
11			26		
12			27		
13			28		
14			29		
15			30		
			31		

Start weight/BMI/BF% Alt metric: _____	Week 1	Week 2	Week 3	Week 4

Start weight/BMI/BF% Alt metric: _____	Week 1	Week 2	Week 3	Week 4

FIGURE 3–5. Accountability chart

the case that you are not interested in weight but are interested in performance (e.g., mile run), which can be used as an alternative metric.

Each day, all you have to do is put a checkmark or an "x" for each goal, indicating that you either did the goal or did not. At the end of the week, you should have between 0–14 checkmarks for the two goals. If you have 14 checkmarks, or something close, and you did not lose weight, then you are either lying about the checkmarks or the behavior that you set out to do was not impactful enough. For example, if you only have two glasses of sugary drinks each week, then the impact of cutting those two drinks out will be low. If you already run 8 minutes a day, and now you commit to running 10 minutes each day, the impact will be low. If you made goals that did not impact your behavior enough, you would need to make new goals with a higher impact.

The progress you make by using the accountability chart will eventually plateau. If you have made the goals a habit, then it may be time to make two new goals. The new goals can focus on completely different behaviors or be revised as more intense versions of the same goal. The accountability chart is simple and effective. It allows you to see and track the results of your behavior change easily, without having to document everything you eat or every workout set.

Part 3: Micro and Macro

> NFPA 1500: Standard on Fire Department Occupational Safety, Health, and Wellness Program and NFPA 1583: Standard on Health-Related Fitness Programs for Fire Department Members state that nutrition should be a part of a prevention and health promotion program. You can find that language in NFPA 1500 Annex A Explanatory Material (Section A.12.2.1) and in NFPA 1583 Annex A Explanatory Material (Section A.8.1.2).

Nutrition and Heart Attack

Nutrition has a major impact on heart health. In 1958, Dr. Ancel Keys conducted a study that has now become famous worldwide on the effect of diet on heart health.[83] The Seven Countries Study compared diets from the U.S., Finland, Netherlands, Italy, Greece, Croatia, Serbia, and Japan. The study found that the food habits of the Mediterranean countries led to a 39% decrease in mortality from heart issues and a 29% decrease in mortality from blood vessel issues compared to the other regions studied.[84] Interestingly, since the study lasted over 50 years, it was able to identify a trend that as Mediterranean people began to eat more of a "Western" diet and become more sedentary, their rates of heart disease significantly increased.[85]

The CDC has been studying dietary patterns in the U.S. since 1989. The Behavioral Risk Factor Surveillance System (BRFSS) is the survey used to assess dietary patterns—specifically the intake of fruits and vegetables.[86] The latest data showed that only 10% of Americans eat enough vegetables and only 12.3% eat enough fruit.[87] In other words, roughly 90% of Americans are not meeting the required fruit and vegetable recommendations. If calories that should be coming from fresh fruits and vegetables are instead coming from processed foods, your risk of heart disease increases.

A major 15-year study that looked at the correlation between added-sugar intake and heart disease found that your risk of dying from heart disease is very closely correlated to the amount of added sugar in your diet, regardless of your age, sex, activity level, or BMI.[88] Astonishingly, the study explains that if you eat 10%–24.9% of your diet as added sugars, your all-cause mortality rate increases by 30% compared to someone who eats less than 10% of their diet as added sugars. If you eat over 25% of your diet as added sugars, your all-cause mortality risk increases by 275% (2.75 times more likely)!

Eating excessive amounts of processed foods and added sugars increases your risks of comorbidities such as obesity, diabetes, cholesterol imbalances, hypertension, and hardening of your blood vessels, which all significantly increase your chances of a heart attack. Scientists continue to search for medications to combat the chronic diseases that are brought on by poor lifestyle, but a diet of fresh, natural foods seems to be more effective than three of the top drugs being researched to improve health outcomes.[89] The University of Sydney found that a good diet was far more powerful than the effect of the drugs which targeted metabolic health (i.e., metformin, rapamycin, and resveratrol).

Over 73% of adults over 20 years old in the U.S. are either overweight or obese.[90] Genetics play a role in obesity, but they are not the primary factor.[91] Obesity is considered a lifestyle disease with nutritional behavior as a large portion of the equation.[92] Unfortunately, firefighters in the U.S. are not any better off than the general population as a group.[93] A major study that looked at data over the past 70 years found that being overweight increased your chance of stroke, heart attack, heart failure, or death from cardiovascular issues by 21% for middle-aged men and 32% for middle-aged women.[94] Those that were obese had risks that were far higher, at 67% for men and 85% for women.

Nutritional behavior plays a major role in your heart health. Heart attacks are still the number one cause of line-of-duty deaths (LODDs) for the fire service. Don't let heart disease be the reason your firefighting career ends.

Nutrition and Cancer

Nutrition has a major impact on cancer. A team from the MD Anderson Cancer Center at the University of Texas, Houston published a research study with the provocative title, "Cancer Is a Preventable Disease That Requires Major Lifestyle Changes."[95] The study found that only 5%–10% of cancers can be attributed to genetics, and the other 90%–95% of cancers can be attributed to lifestyle. The largest contributor of cancer from lifestyle factors was diet, making up 30%–35%. Tobacco made up 25%–30%, infections made up 15%–20%, obesity made up 10%–20%, and alcohol made up 4%–6%. The remainder was compiled as a group called "other." This research suggests that nutritional behavior (diet, obesity, and alcohol) plays a role in up to 61% of all cancers. The study explains that diet specifically is linked in up to 70% of colorectal cancers. The study listed preservatives, pesticides, and other food additives as concerns.

A National Institute for Occupational Safety and Health (NIOSH) study on firefighter cancer identified 19 cancers associated with firefighting.[96] The study found that firefighters have a 9% increased risk of getting cancer and a 14% increased risk of dying if they get cancer compared to the general public. However, the authors explained in their discussion section that there was an excess number of digestive cancers, mainly of the esophageal and colorectal sites. The authors noted that lifestyle factors such as diet, obesity, physical activity, tobacco, and alcohol seem to play a bigger role in digestive tract cancers compared to occupational hazards, according to the literature. As it turns out, 13 cancers have been associated with increased body weight, and 8 of those cancers overlap with the cancers in the NIOSH study.[97] Those 8 cancers include esophagus, stomach, large intestine (colon), rectum, breast, kidney, brain, and multiple myeloma.

The link between nutrition and cancer goes beyond obesity. It is now known that certain foods actually fight cancer, as shown in the www.eattobeat.org database. It is also known that the foods you eat can impact major systems within your body such as your microbiome, DNA, stem cells, immunity, and angiogenesis, all of which play a role in your body's fight against cancer.[98] One easy habit to adopt is eating blueberries—they increase your immune cells by 88% after 6 weeks.[99] Another easy habit is switching from milk chocolate to dark chocolate (70% cacao or higher), which will help your body produce up to twice as many stem cells.[100] The same is true about red wine compared to beer or vodka.[101] The science on how certain foods help your body continues to grow. Dr. Li's book, *Eat to Beat Disease,* provides an excellent overview.

Food is not just calories—it is information. Your nutritional habits guide your body's direction at a cellular level. Cancer is the second leading cause of death, for all ages, in the U.S. and is a major concern for firefighters.[102] Recall from chapter 1 that when you constrict the mortality statistics to the closest age range we can get to the working age for firefighters (15–64 years old), cancer becomes the leading cause of death. Obesity plays a major role, but so does the type of food you eat. The more preservatives, additives, and pesticides you ingest, the greater your cancer risk. With the information you have now, every meal you eat can help you to fight cancer.

Nutrition and Suicide

Nutrition has a major impact on mental health. There are different reasons why someone may die by suicide. However, psychological autopsies have shown that roughly 90% of suicide victims had a treatable mental disorder.[103] It has also been shown that depression and anxiety are both major risk factors for suicide.[104] In fact, one study found that almost half of all suicide deaths around the world can be attributed to depression.[105] Another study found that over 70% of individuals who reported at least one suicide attempt in their life had an anxiety disorder.[106] You should understand that nutritional behavior can impact your risk for depression and anxiety, and thus suicide.

There is a growing field of science called nutritional psychology that studies the connection between what you eat and how you think and feel. A major focus of the nutritional psychology field is on the activity in your microbiome. The reason for studying this area is that your microbiome produces neurotransmitters such as serotonin, dopamine, and GABA, all of which affect your mood.[107] A deficiency in any of these three major neurotransmitters can increase your risk for depression and anxiety. A meta-analysis involving many different dietary patterns found that those who adhere to a Mediterranean diet had a 33% lower risk of depression.[108] One reason for this is because the Mediterranean diet offers more nutrients than the standard western diet.[109] A second reason is that it is healthier for your microbiome.[110]

One of the main hallmarks of the Western "diet" is the large amount of processed food and added sugars and fats. There appears to be a positive correlation between processed food and rates of depression. In a study that looked at fast food consumption, those who ate the most fast food had a 40% higher risk of depression than those who ate the least.[111] Another study found that men who consumed the most sugar had a 23% increased risk of depression.[112] Furthermore, compared to those who did not drink sugar sweetened beverages, those who have the equivalent of three or more sodas a day have a 25% higher risk of depression.[113] Those beverages might not be soda—they could also be sweet tea, sweetened coffee, sports drinks, or energy drinks. As it turns out, the most common source of additional sugars for Americans comes from sugary drinks.[114]

Carrying extra body fat negatively impacts your brain function. According to the CDC, the obesity rate in the U.S. was only 12% in 1991.[115] Now, it is over 40%.[116] It makes sense then that obesity has been classified as an epidemic by both the CDC and the World Health Organization (WHO). If you have more body fat, especially around the waist, you likely have lower gray matter in your brain.[117] Lower gray matter can lower your cognitive function.[118] It can also negatively impact your mood.[119] However, there are other brain-related risks when you put on excess weight, such as the increased likelihood of mental disorders. In fact, being obese increases your risk of mental disorders anywhere from 25% up to 55%.[120] The reason appears to be the inflammation in the brain and the body that is caused by poor dietary choices such as excess sugar and excess dietary fat intake.[121]

Nutritional wellness is a major factor of your mental health. Your ability to think clearly and control your emotions impacts your job as a firefighter and can help prevent suicidal tendencies. Fuel your brain with healthy food.

Conclusion

1. You cannot change the nutrients your body needs, the way they work, or their impact.
2. You can change the variety, the density, and the quantity of nutrients that you consume.
3. Optimizing nutritional behavior might include the following behaviors:
    a. Balancing your nutrient variety using the MyPlate model.
    b. Choosing foods that fight cancer and have a high nutrient density and low GI score.
    c. Making goals that focus on high-impact changes to your nutritional behaviors.
4. Your nutritional behaviors have a direct effect on body systems, organs, and cells.
5. The health of your body systems, organs, and cells impacts your risk of heart attack, cancer, and suicide.

Heart attack, cancer, and suicide are health issues that are killing firefighters more than workplace trauma. Fire departments must take active measures to educate their members on how to optimize health. Firefighters must then make wellness choices that optimize their health. Until these two efforts are being done, demands for support regarding heart attack, cancer, and suicide will not be taken completely seriously by legislators and the public.

Your body relies on fuel to function. The fuel you put in your body comes from your nutritional choices. The fuel you choose provides information to your body and brain, and that information is different based on the variety of the food you eat, the quality of the food you eat, and the quantity of the food you eat. The fuel you put in your body can help to optimize or de-optimize your functionality not just on the fireground, but also in your ability to fight cancer, control your emotions, think clearly, and keep mental disorders in check.

End of Chapter Questions

1. What are three things you cannot change about nutrition?
2. What are three things you can change about nutrition?
3. What are the two different classifications of nutrients?
4. What are the guidelines for the MyPlate model?
5. Why is it important to eat different colors of food?
6. What is the equation for nutrient density?
7. In what ways can food be processed?
8. Identify at least three considerations for how food is raised.
9. What is the USDA's definition of organic?
10. What agricultural product did the International Agency for Research on Cancer (IARC) label as a probable carcinogen to humans?
11. Where can you find a list of foods with the most pesticides on them?
12. What is the Glycemic Index?
13. What is the Aggregate Nutrient Density Index?
14. What is the Eat to Beat Cancer database?
15. What is the easy way to calculate how many calories you need for your desired weight?
16. What is the easy way to calculate how much water you should drink each day?
17. Roughly how many calories does it take to gain or lose a pound of body weight?
18. Give one example for how nutrition reduces your risk of heart attack, cancer, and suicide.
19. Which section in NFPA 1500 addresses nutrition?
20. Which section in NFPA 1583 addresses nutrition?

Notes

1. Gemma Chiva-Blanch and Lina Badimon, "Benefits and Risks of Moderate Alcohol Consumption on Cardiovascular Disease: Current Findings and Controversies," *Nutrients* 12, no. 1 (2019): 108, https://doi.org/10.3390/nu12010108; Ascensión Marcos et al., "Moderate Consumption of Beer and Its Effects on Cardiovascular and Metabolic Health: An Updated Review of Recent Scientific Evidence," *Nutrients* 13, no. 3 (2021): 879, https://doi.org/10.3390/

nu13030879; Kerstin Damström Thakker, "An Overview of Health Risks and Benefits of Alcohol Consumption," *Alcoholism: Clinical and Experimental Research* 22 (1998): 285s–98s; Elizabeth Mostofsky et al., "Key Findings on Alcohol Consumption and a Variety of Health Outcomes from the Nurses' Health Study," *American Journal of Public Health* 106, no. 9 (2016): 1586–91, https://doi.org/10.2105/AJPH.2016.303336.
2. Terry A. Lennie et al., "Micronutrient Deficiency Independently Predicts Time to Event in Patients with Heart Failure," *Journal of the American Heart Association* 7, no. 17 (2018): e007251, https://doi.org/10.1161/JAHA.117.007251; Edoardo Sciatti et al., "Nutritional Deficiency in Patients with Heart Failure," *Nutrients* 8, no. 7 (2016): 442, https://doi.org/10.3390/nu8070442; Nils Bomer et al., "Micronutrient Deficiencies in Heart Failure: Mitochondrial Dysfunction as a Common Pathophysiological Mechanism?" *Journal of Internal Medicine* 291, no. 6 (2022): 713–31, https://doi.org/10.1111/joim.13456.
3. Uwe Gröber et al., "Micronutrients in Oncological Intervention," *Nutrients* 8, no. 3 (2016): 163, https://doi.org/10.3390/nu8030163; Bruce N. Ames and Patricia Wakimoto, "Are Vitamin and Mineral Deficiencies a Major Cancer Risk?" *Nature Reviews Cancer* 2, no. 9 (2002): 694–704, https://doi.org/10.1038/nrc886; Bruce N. Ames, "Micronutrients Prevent Cancer and Delay Aging," *Toxicology Letters* 102 (1998): 5–18, https://doi.org/10.1016/s0378-4274(98)00269-0.
4. T. S. Sathyanarayana Rao et al., "Understanding Nutrition, Depression and Mental Illnesses," *Indian Journal of Psychiatry* 50, no. 2 (2008): 77–82, https://doi.org/10.4103/0019-5545.42391; Maurizio Muscaritoli, "The Impact of Nutrients on Mental Health and Well-Being: Insights from the Literature," *Frontiers in Nutrition* (2021): 97, https://doi.org/10.3389/fnut.2021.656290; Penny M. Kris-Etherton et al., "Nutrition and Behavioral Health Disorders: Depression and Anxiety," *Nutrition Reviews* 79, no. 3 (2021): 247–60, https://doi.org/10.1093/nutrit/nuaa025.
5. Edward F. Coyle, "Fat Metabolism During Exercise: New Concepts," *Sports Science Exchange* 59, no. 8 (1995): 6, https://www.gssiweb.org/en-ca/article/sse-59-fat-metabolism-during-exercise-new-concepts.
6. Tatiana El Bacha, Maurício R. M. P. Luz, and Andrea T. Da Poian, "Dynamic Adaptation of Nutrient Utilization in Humans," *Nature Education* 3, no. 9 (2010): 8.
7. Cedric X. Bryant, Daniel J. Green, and Sabrena Newton-Merrill, *Ace Health Coach Manual: The Ultimate Guide to Wellness, Fitness, & Lifestyle Change* (American Council on Exercise, 2013).
8. Ghada A. Soliman, "Dietary Fiber, Atherosclerosis, and Cardiovascular Disease," *Nutrients* 11, no. 5 (2019): 1155, https://doi.org/10.3390/nu11051155.
9. Satish Sanku Chander Rao, S. Yu, and A. Fedewa, "Systematic Review: Dietary Fibre and FODMAP-Restricted Diet in the Management of Constipation and Irritable Bowel Syndrome," *Alimentary Pharmacology & Therapeutics* 41, no. 12 (2015): 1256–70.

10. Kassem Makki et al., "The Impact of Dietary Fiber on Gut Microbiota in Host Health and Disease," *Cell Host & Microbe* 23, no. 6 (2018): 705–15, https://doi.org/10.1016/j.chom.2018.05.012.
11. Bryant, Green, and Newton-Merrill, *Ace Health Coach Manual*.
12. Micah Craig, Siva Naga S. Yarrarapu, and Manjari Dimri, "Biochemistry, Cholesterol," StatPearls Publishing, last modified August 8, 2018, https://www.ncbi.nlm.nih.gov/books/NBK513326/.
13. Walter L. Miller and Richard J. Auchus, "The Molecular Biology, Biochemistry, and Physiology of Human Steroidogenesis and Its Disorders," *Endocrine Reviews* 32, no. 1 (2011): 81–151, https://doi.org/10.1210/er.2010-0013.
14. Bryant, Green, and Newton-Merrill, *Ace Health Coach Manual*.
15. Bryant, Green, and Newton-Merrill, *Ace Health Coach Manual*.
16. Stephen Nussey and Saffron Whitehead, "Principles of Endocrinology," in *Endocrinology: An Integrated Approach*, ed. Stephen Nussey and Saffron Whitehead (BIOS Scientific Publishers, 2001), https://www.ncbi.nlm.nih.gov/books/NBK20/.
17. Frank Q. Nuttall and Mary C. Gannon, "Dietary Protein and the Blood Glucose Concentration," *Diabetes* 62, no. 5 (2013): 1371–72, https://doi.org/10.2337/db12-1829; George A. Soultoukis and Linda Partridge, "Dietary Protein, Metabolism, and Aging," *Annual Review of Biochemistry* 85 (2016): 5–34.
18. Yves Schutz, "Role of Substrate Utilization and Thermogenesis on Body-Weight Control with Particular Reference to Alcohol," *Proceedings of the Nutrition Society* 59, no. 4 (2000): 511–17, https://doi.org/10.1017/s0029665100000744.
19. Eric B. Rimm et al., "Review of Moderate Alcohol Consumption and Reduced Risk of Coronary Heart Disease: Is the Effect Due to Beer, Wine, or Spirits?" *BMJ* 312, no. 7033 (1996): 731–36; Anna G. Hoek et al., "Alcohol Consumption and Cardiovascular Disease Risk: Placing New Data in Context," *Current Atherosclerosis Reports* 24, no. 1 (2022): 51–59; Séverine Sabia et al., "Alcohol Consumption and Risk of Dementia: 23 Year Follow-Up of Whitehall II Cohort Study," *BMJ* 362 (2018): 362, https://doi.org/10.1136/bmj.k2927.
20. Joao Tomé-Carneiro et al., "Resveratrol and Clinical Trials: The Crossroad from In Vitro Studies to Human Evidence," *Current Pharmaceutical Design* 19, no. 34 (2013): 6064–93, https://doi.org/10.2174/13816128113199990407.
21. Tomé-Carneiro et al., "Resveratrol and Clinical Trials"; Seth D. Crockett et al., "Inverse Relationship Between Moderate Alcohol Intake and Rectal Cancer: Analysis of the North Carolina Colon Cancer Study," *Diseases of the Colon and Rectum* 54, no. 7 (2011): 887–94, https://doi.org/10.1007/DCR.0b013e3182125577; Shan He, Cuirong Sun, and Yuanjiang Pan, "Red Wine Polyphenols for Cancer Prevention" *International Journal of Molecular Sciences* 9, no. 5 (2008): 842–53, https://doi.org/10.3390/ijms9050842.
22. Coyle, "Fat Metabolism During Exercise."
23. Carlos Celis-Morales et al., "Physical Activity Attenuates the Effect of the FTO Genotype on Obesity Traits in European Adults: The Food4Me Study," *Obesity* 24, no. 4: 962–69, https://doi.org/10.1002/oby.21422; Ruth McPherson, "Genetic

Contributors to Obesity," *Canadian Journal of Cardiology* 23 (2007): 23A–27A, https://doi.org/10.1016/s0828-282x(07)71002-4; Vidhu V. Thaker, "Genetic and Epigenetic Causes of Obesity," *Adolescent Medicine: State of the Art Reviews* 28, no. 2 (2017): 379–405; Xiang Li, and Lu Qi, "Gene–Environment Interactions on Body Fat Distribution," *International Journal of Molecular Sciences* 20, no. 15 (2019): 3690, https://doi.org/10.3390/ijms20153690.

24. Bryant, Green, and Newton-Merrill, *Ace Health Coach Manual*.
25. Bryant, Green, and Newton-Merrill, *Ace Health Coach Manual*; Anne-Laure Tardy et al., "Vitamins and Minerals for Energy, Fatigue and Cognition: A Narrative Review of the Biochemical and Clinical Evidence," *Nutrients* 12, no. 1 (2020): 228, https://doi.org/10.3390/nu12010228; Edward Huskisson, Silvia Maggini, and Michael Ruf, "The Role of Vitamins and Minerals in Energy Metabolism and Well-Being," *Journal of International Medical Research* 35, no. 3 (2007): 277–89, https://doi.org/10.1177/147323000703500301.
26. Bryant, Green, and Newton-Merrill, *Ace Health Coach Manual*; Bruce N. Ames, "DNA Damage from Micronutrient Deficiencies Is Likely to Be a Major Cause of Cancer," *Mutation Research/Fundamental and Molecular Mechanisms of Mutagenesis* 475, no. 1–2 (2001): 7–20, https://doi.org/10.1016/s0027-5107(01)00070-7; Ryan Janjuha et al., "Effects of Dietary or Supplementary Micronutrients on Sex Hormones and IGF-1 in Middle and Older Age: A Systematic Review and Meta-Analysis," *Nutrients* 12, no. 5 (2020): 1457, https://doi.org/10.3390/nu12051457.
27. Bryant, Green, and Newton-Merrill, *Ace Health Coach Manual*.
28. Zewen Liu et al., "Role of ROS and Nutritional Antioxidants in Human Diseases," *Frontiers in Physiology* 9 (2018): 477, https://doi.org/10.3389/fphys.2018.00477; Bee Ling Tan et al., "Antioxidant and Oxidative Stress: A Mutual Interplay in Age-Related Diseases," *Frontiers in Pharmacology* 9 (2018): 1162, https://doi.org/10.3389/fphar.2018.01162; Carmia Borek, "Dietary Antioxidants and Human Cancer," *Integrative Cancer Therapies* 3, no. 4 (2004): 333–41, https://doi.org/10.1177/1534735404270578; Alok Ranjan et al., "Role of Phytochemicals in Cancer Prevention," *International Journal of Molecular Sciences* 20, no. 20 (2019): 4981, https://doi.org/10.3390/ijms20204981.
29. Bryant, Green, and Newton-Merrill, *Ace Health Coach Manual*; Barry M. Popkin, Kristen E. D'Anci, and Irwin H. Rosenberg, "Water, Hydration, and Health," *Nutrition Reviews* 68, no. 8 (2010): 439–58, https://doi.org/10.1111/j.1753-4887.2010.00304.x.
30. Bryant, Green, and Newton-Merrill, *Ace Health Coach Manual*; Popkin, D'Anci, and Rosenberg, "Water, Hydration, and Health."
31. Bryant, Green, and Newton-Merrill, *Ace Health Coach Manual*.
32. Valentina Tremaroli and Fredrik Bäckhed, "Functional Interactions Between the Gut Microbiota and Host Metabolism," *Nature* 489, no. 7415 (2012): 242–49.
33. M. J. Hill, "Intestinal Flora and Endogenous Vitamin Synthesis," *European Journal of Cancer Prevention* 6, no. 2 (1997): S43–45, https://doi.org/10.1097/00008469-199703001-00009.

34. Danping Zheng, Timur Liwinski, and Eran Elinav, "Interaction Between Microbiota and Immunity in Health and Disease," *Cell Research* 30, no. 6 (2020): 492–506, https://doi.org/10.1038/s41422-020-0332-7.
35. Megan Clapp et al., "Gut Microbiota's Effect on Mental Health: The Gut-Brain Axis," *Clinics and Practice* 7, no. 4 (2017): 987, https://doi.org/10.4081/cp.2017.987; Philip Strandwitz, "Neurotransmitter Modulation by the Gut Microbiota," *Brain Research* 1693 (2018): 128–33, https://doi.org/10.1016/j.brainres.2018.03.015; Yijing Chen, Jinying Xu, and Yu Chen, "Regulation of Neurotransmitters by the Gut Microbiota and Effects on Cognition in Neurological Disorders," *Nutrients* 13, no. 6 (2021): 2099, https://doi.org/10.3390/nu13062099.
36. Mark L. Heiman and Frank L. Greenway, "A Healthy Gastrointestinal Microbiome Is Dependent on Dietary Diversity," *Molecular Metabolism* 5, no. 5 (2016): 317–20, https://doi.org/10.1016/j.molmet.2016.02.005.
37. Julia Kaźmierczak-Barańska, Karolina Boguszewska, and Boleslaw T. Karwowski, "Nutrition Can Help DNA Repair in the Case of Aging," *Nutrients* 12, no. 11 (2020): 3364, https://doi.org/10.3390/nu12113364; J. C. Mathers, J. M. Coxhead, and J. Tyson, "Nutrition and DNA Repair—Potential Molecular Mechanisms of Action," *Current Cancer Drug Targets* 7, no. 5 (2007): 425–31, https://doi.org/10.2174/156800907781386588.
38. Tabitha M. Hardy and Trygve O. Tollefsbol, "Epigenetic Diet: Impact on the Epigenome and Cancer," *Epigenomics* 3, no. 4 (2011): 503–18, https://doi.org/10.2217/epi.11.71.
39. Serena Galiè et al., "Impact of Nutrition on Telomere Health: Systematic Review of Observational Cohort Studies and Randomized Clinical Trials," *Advances in Nutrition* 11, no. 3 (2020): 576–601, https://doi.org/10.1093/advances/nmz107; Silvia Canudas et al., "Mediterranean Diet and Telomere Length: A Systematic Review and Meta-Analysis," *Advances in Nutrition* 11, no. 6 (2020): 1544–54, https://doi.org/10.1093/advances/nmaa079; Ligi Paul, "Diet, Nutrition and Telomere Length," *The Journal of Nutritional Biochemistry* 22, no. 10 (2011): 895–901, https://doi.org/10.1016/j.jnutbio.2010.12.001.
40. Wojciech Zakrzewski et al., "Stem Cells: Past, Present, and Future," *Stem Cell Research & Therapy* 10, no. 1 (2019): 1–22, https://doi.org/10.1186/s13287-019-1165-5.
41. Ron Sender and Ron Milo, "The Distribution of Cellular Turnover in the Human Body," *Nature Medicine* 27, no. 1 (2021): 45–48, https://doi.org/10.1038/s41591-020-01182-9; Li, *Eat to Beat Disease*.
42. William W. Li, *Eat to Beat Disease: The New Science of How Your Body Can Heal Itself* (Grand Central Publishing, 2019).
43. Jean S. Marshall et al., "An Introduction to Immunology and Immunopathology," *Allergy, Asthma & Clinical Immunology* 14, no. 2 (2018): 49, https://doi.org/10.1186/s13223-018-0278-1; Institute for Quality and Efficiency in Health Care, "The Innate and Adaptive Immune Systems," InformedHealth.org, last modified July 30, 2020, https://www.ncbi.nlm.nih.gov/books/NBK279396/.

44. Camilla Barbero Mazzucca et al., "How to Tackle the Relationship Between Autoimmune Diseases and Diet: Well Begun Is Half-Done," *Nutrients* 13, no. 11 (2021): 3956, https://doi.org/10.3390/nu13113956; Carina Venter et al., "Nutrition and the Immune System: A Complicated Tango," *Nutrients* 12, no. 3 (2020): 818, https://doi.org/10.3390/nu12030818; Caroline E. Childs, Philip C. Calder, and Elizabeth A. Miles, "Diet and Immune Function," *Nutrients* 11, no. 8 (2019): 1933, https://doi.org/10.3390/nu11081933.
45. Brittany A. Potz et al., "Novel Molecular Targets for Coronary Angiogenesis and Ischemic Heart Disease," *Coronary Artery Disease* 28, no. 7 (2017): 605–13, https://doi.org/10.1097/MCA.0000000000000516; Thomas H. Adair and Jean-Pierre Montani, "Chapter 1: Overview of Angiogenesis," in *Angiogenesis* (Morgan & Claypool Life Sciences, 2010), https://www.ncbi.nlm.nih.gov/books/NBK53238/; Peng Li et al., "PubAngioGen: A Database and Knowledge for Angiogenesis and Related Diseases," *Nucleic Acids Research* 43, no. D1 (2015): D963–67, https://doi.org/10.1093/nar/gku1139; Priya Nijhawans, Tapan Behl, and Shaveta Bhardwaj, "Angiogenesis in Obesity," *Biomedicine & Pharmacotherapy* 126 (2020): 110103, https://doi.org/10.1016/j.biopha.2020.110103; H. Roger Lijnen, "Angiogenesis and Obesity," *Cardiovascular Research* 78, no. 2 (2008): 286–93, https://doi.org/10.1093/cvr/cvm007.
46. "MyPlate," U.S. Department of Agriculture, accessed July 18, 2023, https://www.myplate.gov/.
47. "Mediterranean Diet Is Best Diet—Once Again," *Scripps*, January 3, 2023, https://www.scripps.org/news_items/6276-mediterranean-diet-is-best-diet-once-again; Giuseppina Augimeri and Daniela Bonofiglio, "The Mediterranean Diet as a Source of Natural Compounds: Does It Represent a Protective Choice Against Cancer?" *Pharmaceuticals* 14, no. 9 (2021): 920, https://doi.org/10.3390/ph14090920; Antonio Ventriglio et al., "Mediterranean Diet and Its Benefits on Health and Mental Health: A Literature Review," *Clinical Practice and Epidemiology in Mental Health* 16, Suppl. 1 (2020): 156–64, https://doi.org/10.2174/1745017902016010156.
48. Philip J. Brooks et al., "The Alcohol Flushing Response: An Unrecognized Risk Factor for Esophageal Cancer from Alcohol Consumption," *PLOS Medicine* 6, no. 3 (2009): e1000050, https://doi.org/10.1371/journal.pmed.1000050.
49. Bo Xi et al., "Relationship of Alcohol Consumption to All-Cause, Cardiovascular, and Cancer-Related Mortality in U.S. Adults," *Journal of the American College of Cardiology* 70, no. 8 (2017): 913–22, https://doi.org/10.1016/j.jacc.2017.06.054.
50. Melinda M. Manore, "Exercise and the Institute of Medicine Recommendations for Nutrition," *Current Sports Medicine Reports* 4, no. 4 (2005): 193–98, https://doi.org/10.1097/01.csmr.0000306206.72186.00.
51. Deanna M. Minich, "A Review of the Science of Colorful, Plant-Based Food and Practical Strategies for 'Eating the Rainbow,'" *Journal of Nutrition and Metabolism* 2019 (2019): 2125070, https://doi.org/10.1155/2019/2125070; Ezgi Doğan Cömert, Burçe Ataç Mogol, and Vural Gökmen, "Relationship Between

Color and Antioxidant Capacity of Fruits and Vegetables," *Current Research in Food Science* 2 (2020): 1-10, https://doi.org/10.1016/j.crfs.2019.11.001.
52. Katherine D. McManus, "Phytonutrients: Paint Your Plate with the Colors of the Rainbow," *Harvard Health Publishing*, April 25, 2019, https://www.health.harvard.edu/blog/phytonutrients-paint-your-plate-with-the-colors-of-the-rainbow-2019042516501.
53. Ting-Ting Huang et al., "Current Understanding of Gut Microbiota in Mood Disorders: An Update of Human Studies," *Frontiers in Genetics* 10 (2019): 98.
54. Hill, "Intestinal Flora and Endogenous Vitamin Synthesis."
55. Zheng, Liwinski, and Elinav, "Interaction Between Microbiota and Immunity."
56. Simon Carding et al., "Dysbiosis of the Gut Microbiota in Disease," *Microbial Ecology in Health and Disease* 26, no. 1 (2015): 26191, https://doi.org/10.3402/mehd.v26.26191.
57. Joel Fuhrman, *Eat to Live: The Amazing Nutrient-Rich Program for Fast and Sustained Weight Loss*, rev. ed. (Little, Brown Spark, 2011).
58. Bernard Srour et al., "Ultra-Processed Food Intake and Risk of Cardiovascular Disease: Prospective Cohort Study (NutriNet-Santé)," *BMJ* 365 (2019): l1451, https://doi.org/10.1136/bmj.l1451.
59. Thibault Fiolet et al., "Consumption of Ultra-processed Foods and Cancer Risk: Results from NutriNet-Santé Prospective Cohort," *BMJ* 360 (2018): k322, https://doi.org/10.1136/bmj.k322.
60. "Use of the Term Natural on Food Labeling," U.S. Food & Drug Administration, last modified October 22, 2018, https://www.fda.gov/food/food-labeling-nutrition/use-term-natural-food-labeling.
61. "GMO FAQs," Non-GMO Project, accessed August 15, 2023, https://www.nongmoproject.org/gmo-faq/.
62. "Gluten-Free Labeling of Foods," U.S. Food & Drug Administration, last modified March 7, 2022, https://www.fda.gov/food/food-labeling-nutrition/gluten-free-labeling-foods.
63. "'Free Range' and 'Pasture Raised' Officially Defined by HFAC for Certified Humane® Label," Certified Humane, last modified January 16, 2014, https://certifiedhumane.org/free-range-and-pasture-raised-officially-defined-by-hfac-for-certified-humane-label/.
64. Oscar, "What Does Cage Free Mean?" *Greener Choices*, February 16, 2021, https://www.greenerchoices.org/cage-free-mean/.
65. Oscar, "What Does Pasture Raised Mean?" *Greener Choices*, February 19, 2021, https://www.greenerchoices.org/pasture-raised/.
66. "Overview," Certified Humane, 2016, https://certifiedhumane.org/overview/.
67. "Introduction to Organic," USDA Agriculture Marketing Service, U.S. Department of Agriculture, accessed August 15, 2023, https://www.ams.usda.gov/services/organic-certification/organic-basics.
68. "IARC Monograph on Glyphosate," *International Agency for Research on Cancer*, 2019, https://www.iarc.who.int/featured-news/media-centre-iarc-news-glyphosate/.

69. "Bayer Acknowledges 'Bumps' in $11 Billion Roundup Deal After Judge Raises Doubts," *Reuters*, August 27, 2020, https://www.reuters.com/article/us-bayer-lawsuit/bayer-acknowledges-bumps-in-11-billion-roundup-deal-after-judge-raises-doubts-idUSKBN25N2YK.
70. "Glyphosate," U.S. Environmental Protection Agency, last modified September 23, 2022, https://www.epa.gov/ingredients-used-pesticide-products/glyphosate.
71. Eva Novotny, "Glyphosate, Roundup and the Failures of Regulatory Assessment," *Toxics* 10, no. 6 (2022): 321, https://doi.org/10.3390/toxics10060321.
72. Sonia Vega-López, Bernard J. Venn, and Joanne L. Slavin, "Relevance of the Glycemic Index and Glycemic Load for Body Weight, Diabetes, and Cardiovascular Disease," *Nutrients* 10, no. 10 (2018): 1361, https://doi.org/10.3390/nu10101361; Kevin D. Hall et al., "The Energy Balance Model of Obesity: Beyond Calories in, Calories Out," *The American Journal of Clinical Nutrition* 115, no. 5 (2022): 1243–54, https://doi.org/10.1093/ajcn/nqac031.
73. Dionysios Vlachos et al., "Glycemic Index (GI) or Glycemic Load (GL) and Dietary Interventions for Optimizing Postprandial Hyperglycemia in Patients with T2 Diabetes: A Review," *Nutrients* 12, no. 6 (2020): 1561, https://doi.org/10.3390/nu12061561.
74. Alok Kumar Dwivedi et al., "Associations of Glycemic Index and Glycemic Load with Cardiovascular Disease: Updated Evidence from Meta-Analysis and Cohort Studies," *Current Cardiology Reports* 24, no. 3 (2022): 141–61, https://doi.org/10.1007/s11886-022-01635-2.
75. David J. A. Jenkins et al., "Glycemic Index, Glycemic Load, and Cardiovascular Disease and Mortality," *New England Journal of Medicine* 384, no. 14 (2021): 1312–22, https://doi.org/10.1056/NEJMoa2007123.
76. Joel Fuhrman, "Knowing a Food's Nutrient Density Is Key to Making Good Choices," *Dr. Fuhrman*, November 3, 2022, https://www.drfuhrman.com/blog/238/knowing-a-foods-nutrient-density-is-key-to-making-good-choices.
77. Division of Nutrition, Physical Activity, and Obesity and National Center for Chronic Disease Prevention and Health Promotion, "Adult Obesity Facts," Centers for Disease Control and Prevention, last modified May 17, 2022, https://www.cdc.gov/obesity/data/adult.html.
78. Kathleen Y. Wolin, Kenneth Carson, and Graham A. Colditz, "Obesity and Cancer," *The Oncologist* 15, no. 6 (2010): 556–65, https://doi.org/10.1634/theoncologist.2009-0285; Konstantinos I. Avgerinos et al., "Obesity and Cancer Risk: Emerging Biological Mechanisms and Perspectives," *Metabolism* 92 (2019): 121–35, https://doi.org/10.1016/j.metabol.2018.11.001.
79. "Calorie Counting Made Easy," *Harvard Health Publishing*, July 11, 2020, https://www.health.harvard.edu/staying-healthy/calorie-counting-made-easy.
80. Bryant, Green, and Newton-Merrill, *Ace Health Coach Manual*.
81. Bryant, Green, and Newton-Merrill, *Ace Health Coach Manual*.
82. "Statement from FDA Commissioner Scott Gottlieb, M.D., on the Agency's New Efforts to Strengthen Regulation of Dietary Supplements by Modernizing and

Reforming FDA's Oversight," U.S. Food and Drug Administration, last modified February 11, 2019, https://www.fda.gov/news-events/press-announcements/statement-fda-commissioner-scott-gottlieb-md-agencys-new-efforts-strengthen-regulation-dietary.

83. Seven Countries Study, accessed July 18, 2023, https://www.sevencountriesstudy.com/.
84. Kim T. B. Knoops et al., "Mediterranean Diet, Lifestyle Factors, and 10-Year Mortality in Elderly European Men and Women: The HALE Project," *JAMA* 292, no. 12 (2004): 1433–39, https://doi.org/10.1001/jama.292.12.1433.
85. Anthony Kafatos et al., "Heart Disease Risk-Factor Status and Dietary Changes in the Cretan Population Over the Past 30 Y: The Seven Countries Study," *The American Journal of Clinical Nutrition* 65, no. 6 (1997): 1882–86, https://doi.org/10.1093/ajcn/65.6.1882.
86. National Center for Chronic Disease Prevention and Health Promotion, Division of Population Health, "About BRFSS," Centers for Disease Control and Prevention," last modified May 16, 2014, https://www.cdc.gov/brfss/about/index.htm.
87. Seung Hee Lee et al., "Adults Meeting Fruit and Vegetable Intake Recommendations—United States, 2019," *Morbidity and Mortality Weekly Report* 71, no. 1 (2022): 1–9, http://dx.doi.org/10.15585/mmwr.mm7101a1.
88. Quanhe Yang et al., "Added Sugar Intake and Cardiovascular Diseases Mortality Among US Adults," *JAMA Internal Medicine* 174, no. 4 (2014): 516–24, https://doi.org/10.1001/jamainternmed.2013.13563.
89. David G. Le Couteur et al., "Nutritional Reprogramming of Mouse Liver Proteome Is Dampened by Metformin, Resveratrol, and Rapamycin," *Cell Metabolism* 33, no. 12 (2021): 2367–79, https://doi.org/10.1016/j.cmet.2021.10.016.
90. Kristen Monaco, "Over 73% of U.S. Adults Overweight or Obese," *MedPage Today*, December 11, 2020, https://www.medpagetoday.com/primarycare/obesity/90142; CDC/National Center for Health Statistics, "National Health and Nutrition Examination Survey: 2017–2018 Questionnaire Data—Continuous NHANES," Centers for Disease Control and Prevention, last modified 2020, https://wwwn.cdc.gov/nchs/nhanes/search/datapage.aspx?Component=Questionnaire&Cycle=2017-2018.
91. Division of Nutrition, Physical Activity, and Obesity and National Center for Chronic Disease Prevention and Health Promotion, "Causes of Obesity," Centers for Disease Control and Prevention, last modified March 21, 2022, https://www.cdc.gov/obesity/basics/causes.html?CDC_AA_refVal=https%3A%2F%2Fwww.cdc.gov%2Fobesity%2Fadult%2Fcauses.html; National Center on Birth Defects and Developmental Disabilities and Office of Genomics and Precision Public Health, "Genes and Obesity," Centers for Disease Control and Prevention, last modified May 17, 2013, https://www.cdc.gov/genomics/resources/diseases/obesity/obesedit.htm.

92. James M. Rippe, Suellyn Crossley, and Rhonda Ringer, "Obesity as a Chronic Disease: Modern Medical and Lifestyle Management," *Journal of the American Dietetic Association* 98, no. 10 (1998): S9–15, https://doi.org/10.1016/s0002-8223 (98)00704-4.
93. Michelle Lynn Wilkinson et al., "Peer Reviewed: Physician Weight Recommendations for Overweight and Obese Firefighters, United States, 2011–2012," *Preventing Chronic Disease* 11 (2014): E116, https://doi.org/10.5888/pcd11. 140091; Walker S. C. Poston et al., "The Prevalence of Overweight, Obesity, and Substandard Fitness in a Population-Based Firefighter Cohort," *Journal of Occupational and Environmental Medicine* 53, no. 3 (2011): 266–73, https://doi.org/10.1097/JOM.0b013e31820af362.
94. Sadiya S. Khan et al., "Association of Body Mass Index with Lifetime Risk of Cardiovascular Disease and Compression of Morbidity," *JAMA Cardiology* 3, no. 4 (2018): 280–87, https://doi.org/10.1001/jamacardio.2018.0022.
95. Preetha Anand et al., "Cancer Is a Preventable Disease That Requires Major Lifestyle Changes," *Pharmaceutical Research* 25 (2008): 2097–116, https://doi.org/10.1007/s11095-008-9661-9.
96. Robert D. Daniels et al., "Mortality and Cancer Incidence in a Pooled Cohort of U.S. Firefighters from San Francisco, Chicago and Philadelphia (1950–2009)," *Occupational and Environmental Medicine* 71, no. 6 (2014): 388–97, https://doi.org/10.1136/oemed-2013-101662.
97. Centers for Disease Control and Prevention, "Cancers Associated with Overweight and Obesity Make up 40 Percent of Cancers Diagnosed in the United States," CDC Newsroom, last modified October 3, 2017, https://www.cdc.gov/media/releases/2017/p1003-vs-cancer-obesity.html.
98. Li, *Eat to Beat Disease*.
99. Anand R. Nair et al., "Blueberry Supplementation Attenuates Oxidative Stress Within Monocytes and Modulates Immune Cell Levels in Adults with Metabolic Syndrome: A Randomized, Double-Blind, Placebo-Controlled Trial," *Food & Function* 8, no. 11 (2017): 4118–28, https://doi.org/10.1039/c7fo00815e.
100. Christian Heiss et al., "Improvement of Endothelial Function with Dietary Flavanols Is Associated with Mobilization of Circulating Angiogenic Cells in Patients with Coronary Artery Disease," *Journal of the American College of Cardiology* 56, no. 3 (2010): 218–24, https://doi.org/10.1016/j.jacc.2010.03.039.
101. Po-Hsun Huang et al., "Intake of Red Wine Increases the Number and Functional Capacity of Circulating Endothelial Progenitor Cells by Enhancing Nitric Oxide Bioavailability," *Arteriosclerosis, Thrombosis, and Vascular Biology* 30, no. 4 (2010): 869–77, https://doi.org/10.1161/ATVBAHA.109.200618.
102. CDC/National Center for Health Statistics, "Leading Causes of Death," Centers for Disease Control and Prevention, last modified January 18, 2023, https://www.cdc.gov/nchs/fastats/leading-causes-of-death.htm.

103. Jonathan T. O. Cavanagh et al., "Psychological Autopsy Studies of Suicide: A Systematic Review," *Psychological Medicine* 33, no. 3 (2003): 395–405, https://doi.org/10.1017/s0033291702006943.
104. Silke Bachmann, "Epidemiology of Suicide and the Psychiatric Perspective," *International Journal of Environmental Research and Public Health* 15, no. 7 (2018): 1425, https://doi.org/10.3390/ijerph15071425.
105. Alize J. Ferrari et al., "Burden of Depressive Disorders by Country, Sex, Age, and Year: Findings from the Global Burden of Disease Study 2010," *PLOS Medicine* 10, no. 11 (2013): e1001547, https://doi.org/10.1371/journal.pmed.1001547.
106. Josh Nepon et al., "The Relationship Between Anxiety Disorders and Suicide Attempts: Findings from the National Epidemiologic Survey on Alcohol and Related Conditions," *Depression and Anxiety* 27, no. 9 (2010): 791–98, https://doi.org/10.1002/da.20674.
107. Strandwitz, "Neurotransmitter Modulation by the Gut Microbiota."
108. Camille Lassale et al., "Healthy Dietary Indices and Risk of Depressive Outcomes: A Systematic Review and Meta-Analysis of Observational Studies," *Molecular Psychiatry* 24, no. 7 (2019): 965–86, https://doi.org/10.1038/s41380-018-0237-8.
109. Itandehui Castro-Quezada, Blanca Román-Viñas, and Lluís Serra-Majem, "The Mediterranean Diet and Nutritional Adequacy: A Review," *Nutrients* 6, no. 1 (2014): 231–48, https://doi.org/10.3390/nu6010231.
110. Ravinder Nagpal et al., "Gut Microbiome-Mediterranean Diet Interactions in Improving Host Health," *F1000Research* 8 (2019): 699, https://doi.org/10.12688/f1000research.18992.1; Melisa A. Bailey and Hannah D. Holscher, "Microbiome-Mediated Effects of the Mediterranean Diet on Inflammation," *Advances in Nutrition* 9, no. 3 (2018): 193–206, https://doi.org/10.1093/advances/nmy013; Tarini Shankar Ghosh et al., "Mediterranean Diet Intervention Alters the Gut Microbiome in Older People Reducing Frailty and Improving Health Status: The NU-AGE 1-Year Dietary Intervention Across Five European Countries," *Gut* 69, no. 7 (2020): 1218–28, https://doi.org/10.1136/gutjnl-2019-319654.
111. Almudena Sánchez-Villegas et al., "Fast-Food and Commercial Baked Goods Consumption and the Risk of Depression," *Public Health Nutrition* 15, no. 3 (2012): 424–32, https://doi.org/10.1017/S1368980011001856.
112. Anika Knüppel et al., "Sugar Intake from Sweet Food and Beverages, Common Mental Disorder and Depression: Prospective Findings from the Whitehall II Study," *Scientific Reports* 7, no. 1 (2017): 6287, https://doi.org/10.1038/s41598-017-05649-7.
113. Danqing Hu, Lixiao Cheng, and Wenjie Jiang, "Sugar-Sweetened Beverages Consumption and the Risk of Depression: A Meta-Analysis of Observational Studies," *Journal of Affective Disorders* 245 (2019): 348–55, https://doi.org/10.1016/j.jad.2018.11.015.

114. Adam Drewnowski and Colin D. Rehm, "Consumption of Added Sugars Among US Children and Adults by Food Purchase Location and Food Source," *The American Journal of Clinical Nutrition* 100, no. 3 (2014): 901–7, https://doi.org/10.3945/ajcn.114.089458; Division of Nutrition, Physical Activity, and Obesity and National Center for Chronic Disease Prevention and Health Promotion, "Get the Facts: Sugar-Sweetened Beverages and Consumption," Centers for Disease Control and Prevention, last modified April 11, 2022, https://www.cdc.gov/nutrition/data-statistics/sugar-sweetened-beverages-intake.html.
115. Centers for Disease Control and Prevention, "Obesity Epidemic Increases Dramatically in the United States: CDC Director Calls for National Prevention Effort," CDC Newsroom, last modified October 26, 1999, https://www.cdc.gov/media/pressrel/r991026.htm.
116. Division of Nutrition, Physical Activity, and Obesity and National Center for Chronic Disease Prevention and Health Promotion, "Adult Obesity Facts."
117. Mark Hamer and G. David Batty, "Association of Body Mass Index and Waist-to-Hip Ratio with Brain Structure: UK Biobank Study," *Neurology* 92, no. 6 (2019): e594–600, https://doi.org/10.1212/WNL.0000000000006879.
118. Chelsea M. Stillman et al., "Body–Brain Connections: The Effects of Obesity and Behavioral Interventions on Neurocognitive Aging," *Frontiers in Aging Neuroscience* 9 (2017): 115, https://doi.org/10.3389/fnagi.2017.00115.
119. Lei Li et al., "Gray Matter Volume Alterations in Subjects with Overweight and Obesity: Evidence from a Voxel-Based Meta-Analysis," *Frontiers in Psychiatry* 13 (2022): 955741, https://doi.org/10.3389/fpsyt.2022.955741.
120. Gregory E. Simon et al., "Association Between Obesity and Psychiatric Disorders in the US Adult Population," *Archives of General Psychiatry* 63, no. 7 (2006): 824–30, https://doi.org/10.1001/archpsyc.63.7.824; Genevieve Gariepy, Danit Nitka, and Norbert Schmitz, "The Association Between Obesity and Anxiety Disorders in the Population: A Systematic Review and Meta-Analysis," *International Journal of Obesity* 34, no. 3 (2010): 407–19, https://doi.org/10.1038/ijo.2009.252; Floriana S. Luppino et al., "Overweight, Obesity, and Depression: A Systematic Review and Meta-Analysis of Longitudinal Studies," *Archives of General Psychiatry* 67, no. 3 (2010): 220–29, https://doi.org/10.1001/archgenpsychiatry.2010.2.
121. Stephanie Fulton et al., "The Menace of Obesity to Depression and Anxiety Prevalence," *Trends in Endocrinology & Metabolism* 33, no. 1 (2022): 18–35, https://doi.org/10.1016/j.tem.2021.10.005.

4
MENTAL WELLNESS

Introduction

Mental wellness should be as important to firefighters as their physical wellness. The nature of the job demands it. Firefighters respond to traumatic situations most people will never face. In addition, the always-on-call nature of the job combined with sleep-disrupting shifts prevent the body and mind from relaxing and rejuvenating. Despite these challenges, firefighters must always be ready to perform optimally because other people's lives or property may be in danger.

In addition to improving performance, mental wellness also promotes resilience. This means that firefighters can improve their careers and personal lives by paying attention to their mental wellness. Unfortunately, mental wellness is not given as much attention or training as physical wellness. This chapter will begin with an overview on the biological and psychological factors of mental health. It will then give some evidence-based strategies to improve mental wellness, which in turn will improve your mental health, mental and physical performance, and resilience.

Defining Mental Health
The World Health Organization (WHO) defines mental health as "a state of well-being in which the individual realizes his or her abilities, can cope with normal stresses of life, can work productively and fruitfully, and is able to make a contribution to his or her community."[1] Similarly, in the United States, the Centers for Disease Control and Prevention (CDC) describes mental health as an individual's "emotional, psychological, and social well-being."[2]

Despite the emphasis on well-being, mental health is often used synonymously with mental illness. Someone with "poor mental health" is assumed to be suffering from a mental disorder like clinical anxiety or depression. But this perception is misleading. Although mental health can be positive (psychological well-being) or negative (psychological distress), poor mental health does not automatically translate to a mental disorder. This is because mental health and mental illness are not opposite ends of the same spectrum; rather, they are two different spectrums.

> We all have a mental health and mental illness continuum. As a result, a person with depression (mental disorder) can still have moments of well-being (mental health).

It is also important to note that a mental disorder requires a clinical diagnosis from a medical professional. If the symptoms are minor, not rising to the level of a mental disorder, then an individual has psychological distress. This is a "subclinical" diagnosis that doesn't require medical attention. To help bring clarity to the confusion surrounding mental health, let's break down three concepts which contribute to the idea of overall mental health: psychological distress, psychological well-being, and cognitive functioning.

Psychological distress
Psychological distress is a general term describing a variety of negative psychological symptoms like excessive worry, intense pressure, and emotional strain. It is an adverse response to stressors in your life. Stressors are events or experiences that cause stress. They range from minor inconveniences like being stuck in traffic to major crises like the unexpected death of a spouse. However, stress and stressors are neutral. It is the way we perceive them that determines if they cause distress. At some point, everyone will experience distress—it is just part of being human and shouldn't be a concern. But being in a chronic state of distress is a problem because it can lead to serious physical illnesses like heart disease and cancer. If left unchecked, psychological distress can also develop into a variety of mental disorders like clinical anxiety or depression, which if left unaddressed can increase the risk of suicidal behavior.

Psychological well-being
Unlike psychological distress, psychological well-being is a positive mental state that has two components. The first is called hedonic well-being, which refers to your subjective feelings of happiness. It is both emotional (high positive emotions and low negative emotions) and cognitive (high satisfaction with life). Thus, an individual experiences happiness when positive emotions and satisfaction with life are both high. The second component is called eudaemonic

well-being, which refers to optimal mental wellness efforts. It has six components—self-acceptance, environmental mastery, positive relationships, personal growth, identifying your purpose in life, and autonomy. Essentially, individuals are highly functioning if they feel a strong sense of purpose, can pursue their goals, and have healthy relationships with friends and family. Individuals with high levels of hedonic and eudaemonic well-being are physically healthier, with lower levels of illnesses like heart disease and cancer. In addition, high well-being is a protective factor from mental disorders like clinical anxiety and depression.

Cognitive function

Finally, cognitive function is an umbrella term referring to mental abilities like learning, thinking, reasoning, remembering, problem-solving, decision-making, imagination, perception, and attention. Apart from those with psychological disabilities or impairments, everyone uses their cognitive functioning to navigate the demands of daily life. They are critical for doing your job well. Having lower levels of psychological distress and higher levels of psychological well-being improves cognitive functioning.

Instead of mental health being a vague term that is synonymous with mental illness, think of mental health as a function of these three concepts—psychological distress, psychological well-being, and cognitive functioning (fig. 4–1). Achieving an optimal level of mental health means adjusting what you can

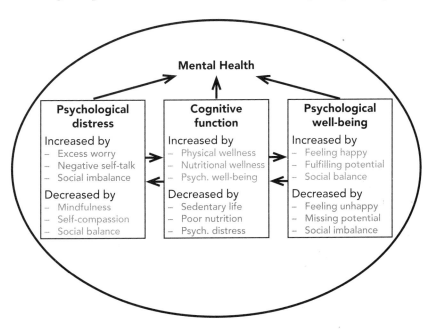

FIGURE 4–1. Components of mental health

about your cognitive functioning so that you minimize psychological distress and maximize psychological well-being. This will then lead to improved cognitive function for areas that you cannot directly adjust. You will learn easier, remember more, make better decisions, solve problems easier, and sharpen your attention.

> Having high levels of well-being translates to a stronger mind, and since whatever happens in the mind is reflected in the brain, it also translates to a stronger brain.

The Brain and the Mind

It is important to understand that mental health involves the brain and mind. While these are often used interchangeably to refer to the same thing, the brain and mind are different. The brain is an organ protected by the skull. It is the epicenter of the nervous system, a collection of billions of interconnected neurons that control and regulate behavior. Since it is physical in nature, scientists can observe the brain using advanced machinery like magnetic resonance imaging (MRI) or computerized tomography (CT scan).

The mind, however, is harder to explain because it is not a physical thing. Rather, it is a word that symbolizes and describes all mental processes like thinking, feeling, remembering, learning, and judging. Since the mind is not a material object, no one can see or touch it. Unlike the brain, machines cannot observe the mind, so studying it requires researchers to rely on people's description of their mental processes. As a result, the information is considered more subjective, making it difficult to generalize concepts about the mind. This does not make the mind any less important than the brain, just more individualized.

There are scientific and philosophical debates about the role of the brain versus the mind, and even whether the mind exists at all. While these are beyond the scope of this chapter, most people agree that the two are inseparable and shape each other. The brain and mind form an interconnected relationship (fig. 4–2). Neural processes in the brain drive the mind's mental processes, and a thought or memory will activate neurons in the brain.

As a result, a change in one reflects a change in the other. For example, researchers found that the mental exercise of learning a second language resulted in structural and functional changes in the brain.[3] After six weeks of language lessons, participants in the study had more integrated neural networks than when they began. In other words, they had stronger connections between the

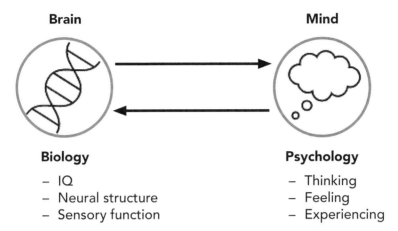

FIGURE 4–2. Relationship between the brain and the mind

different regions of the brain that are active when learning something new. These stronger connections allowed participants to think about the subject matter faster and more efficiently compared to a control group.

This study on language learning and brain development highlights the positive, reciprocal relationship between the brain and mind. The mental challenge of learning a new language made the brain stronger and faster. As a result of the brain's improvements, future learning will be easier. This will allow for more advanced learning, which will further strengthen the brain, and the cycle will continue to repeat itself.

The brain-mind interconnected relationship is central to understanding mental health and performance. But the biological intricacies of the nervous system, combined with the complexities of studying the mind, make understanding this relationship complicated. Fortunately, Dr. Eric Kandel, a Nobel Prize winner and professor at Columbia University specializing in psychiatry and neuroscience, developed a framework describing this relationship.[4]

Dr. Kandel's Brain-Mind Framework

Dr. Kandel's framework has five principles; this section provides a description of each one.

Principle 1: All mental processes reflect brain processes

The most fundamental principle is that the brain and mind function together. According to Dr. Kandel, the actions of the brain drive all human behaviors, from seemingly simple motor behaviors like biting an apple to complex cognitive behaviors like creating a work of art. Whatever we do, whenever we do it, at least part of the brain is active. This means that any time our mind is active, the brain is also active. The two are inseparable.

Principle 2: Genes (and their combinations) control brain functioning, which in turn control behavior, including mental functions and mental illness

Genes are like puzzle pieces; they fit together in a specific way. But instead of creating a picture, the puzzle pieces create directions for how the body operates. In the brain, the combination of genes determines how the brain functions, specifically the structure and strength of the neural connections. Since all behavior involves the brain (Principle 1), genes have a lot of influence on behavior. This is why some people are more susceptible to physical diseases or psychological disorders; the combination of their genes makes them predisposed.

Principle 3: Social, developmental, and environmental factors alter gene expression

However, genes do not explain all behavior. Factors outside of the individual also determine an individual's behavior. Humans are social creatures: we learn from those around us, like parents and friends, and we are also shaped by social and cultural norms. These outside factors can be categorized as (a) social (family, friends), (b) developmental (experiences, memories), and (c) environmental (socioeconomic status, cultural norms). Each one of these categories influences our behavior by teaching us something about the world and, in turn, shaping who we are as people and developing our values and beliefs. The field of epigenetics shows that these outside factors determine gene expression. This means that social, developmental, and environmental factors determine which gene puzzle pieces are selected to create the directions for how the body operates.

Principle 4: Alterations in gene expression induce changes in brain functioning

Our experiences and environment change our biological makeup. By selecting the gene puzzle pieces, the outside factors influence the brain's structure and overall function. Since the brain and mind operate synchronously, and a change in one reflects a change in the other, our unique experiences become the biological basis for individuality.

Principle 5: Learning changes behavior by producing changes in gene expression that alter the strength and structure of the brain

It is important to note that each person responds to the outside factors in different ways. As a result, the changes in the brain-mind relationship occur at the individual level. We learn and adapt from outside factors in different ways, shaping our brains and minds in unique ways. To improve mental health, it is imperative to have experiences and be in environments that create positive change in the brain. The brain-mind relationship is also dynamic, meaning we are always learning something and changing because of it. Everything, from the mundane to the exceptional, leads to structural changes in the brain. Every day, our experiences rewire our neural connections, weakening some and strengthening others. We must be intentional, giving ourselves opportunities to improve our mental wellness efforts. If we don't, then our mental health and performance will stagnate at best or decline at worst.

***

As alluded to throughout the five principles, Dr. Kandel's framework describes mental health as the product of nature and nurture. Principles 1 and 2 emphasize nature, showing how the brain and genes drive all behavior. Principle 3 then illustrates that behavior is reliant on nurture, as our interaction with the environment is responsible for determining gene activation. Finally, Principles 4 and 5 show how nature and nurture work together in a cooperative relationship, influencing the structure and function of the brain.

> It is impossible to separate nature and nurture because they are locked in a dynamic, interdependent dance. As Dr. Kandel remarks, "all 'nurture' is ultimately expressed as 'nature.'"[5]

This chapter will use this simplified version of Dr. Kandel's framework. We'll assume that the brain is involved in all mental processes. This assumption does not mean that biology (study of the brain) is more important than psychology (study of the mind). Rather, anything related to the psychological, even the most complex cognitive concepts or debilitating mental disorders, are not just things that are "in your head" but are inherently biological. Further, due to the interconnected brain-mind relationship, changing one will change the other.

How does this relate to firefighters? The information in this chapter will show how your mental wellness habits—and, as a result, your performance and resilience—are the result of your biology (nature) and your interaction with your environment (nurture). Thus, the strategies to improve your mental wellness habits will depend on modifying your environment and lifestyle within the constraints of your biology.

This chapter is divided into three parts to help guide you through the information. *Part 1: Nature and Nurture* provides an overview of intelligence (IQ) and sensory functions, which are things that you cannot change. It then moves to things that you can change including your thought patterns and how you connect with others. *Part 2: Deficiency and Excess* gives a more thorough description of the biological and psychological components of mental wellness, highlighting the role your choices have on your mental health. In addition, it gives some strategies to improve mental wellness habits and cognitive function. Finally, *Part 3: Micro and Macro* explains how your mental wellness habits impact you at different levels.

Part 1: Nature and Nurture

What You Cannot Change (Nature)

After exploring the basics of mental health, we can now explore what determines mental health. First, mental health is a product of biological and psychological forces, or the result of nature and nurture. Dr. Kandel's mental health framework emphasizes the roles of nature (Principles 1 and 2), nurture (Principle 3), and nature and nurture acting together (Principles 4 and 5). Most of the information and strategies described in this chapter focus on nurture because you can control and change your behavior. Before jumping into that information, however, it is important to understand the underlying genetic (or nature) components of mental health, which you cannot change.

Although genetic research is still in its infancy, researchers are confident that all major mental disorders have some genetic component, ranging from a low of 20%–45% for anxiety and major depression disorders to upwards of 75% for autism, schizophrenia, and ADHD.[6] While these percentages are high, the association between genes and mental disorders isn't straightforward, because no one gene causes any one mental disorder. Instead, multiple genes add or subtract to the risk of developing a mental disorder, which could be triggered by an environmental event or lifestyle behaviors. This is also true of psychological distress.

While you cannot change your genetic predisposition to mental disorders or psychological distress, Dr. Kandel's framework shows that you can lower the chance of developing a mental disorder by changing your behaviors or environments. But change is hard, at least at first, because it involves learning new habits and adapting to new environments. The degree to which people struggle or succeed in changing their behavior is partially influenced by genetics, particularly in two areas—intelligence and sensory function.

Intelligence
Intelligence, typically referred to as IQ, signifies one's cognitive ability, or the ease with which you can process information and solve problems. Higher scores are generally associated with life success, positively predicting outcomes like social status, educational and job performance, health, and longevity. For these reasons, IQ is sometimes considered a taboo topic, but IQ is an important factor in understanding behavior change.

While there is an ongoing debate about the relative impact of genetics compared to environment in determining your IQ, research is bringing some clarity to the discussion. In 2013, a study examined 11,000 twins and found that even though the impact of genetics is much higher than the environmental impact, your environment can influence your IQ to some degree.[7] For those with lower or average IQ, the environmental influence declines rapidly after around 4 years of age. However, for those with high IQ, the environmental influence can extend out to 16 years of age. In other words, the environment plays some role in shaping IQ in childhood, but for most people IQ will be unchangeable by the time they are in elementary school.

Most people associate IQ with the ability to learn and use knowledge. But learning and intelligence are not the same thing. While IQ refers to cognitive ability, learning refers to the process of gaining knowledge. You can think of IQ like your skeleton. Your parents might be able to influence your skeletal size slightly early on in your life by your diet, sleep, and other factors, but only to a certain genetically predetermined point, and that point varies by person.

> Learning is like doing exercise and building muscle. Everyone can benefit from exercise regardless of their genetically predetermined skeletal size. Similarly, learning new information is something that can benefit anyone, which helps to sharpen the brain and mind regardless of IQ.

Sensory function
We have five primary senses—sight, smell, hearing, taste, and touch. The way your five senses function is determined by your genetics. For example, some people are born nearsighted, hindering their ability to see clearly without glasses or contact lenses. Others are born supertasters, making their sense of taste stronger than the average person. While our five senses help us navigate and make sense of the world, they also are tied to the way we learn.

Learning is a multisensory and individual process.[8] In other words, we learn through our five senses. But the way you learn is unique and is at least partly determined by your genetics. This is one of the reasons people prefer different media for learning. What works best for you may not work best for another person.

Despite the individuality of learning, everyone learns better when it is multisensory. Using the full range of senses supercharges learning because it mirrors what our brains do naturally—process information using multiple senses. For example, watching a car drive down the street involves seeing the car, hearing its engine, smelling its exhaust, and tasting the dust it kicks up as it passes. Therefore, unisensory-training that emphasizes only one sense (like listening to a lecture) is less effective because it goes against the way we naturally learn through experiencing the world around us.

The way we process sensory information also varies by person. At the extreme of the sensory continuum is a subset of the population who are classified as Highly Sensitive Persons (HSPs).[9] These people have a personality trait known as sensory-processing sensitivity. This trait makes HSPs display increased emotional sensitivity, which leads to stronger reactions to both external and internal stimuli like pain, hunger, light, and noise. It is important to note that this is not a psychological disorder; rather, HSPs are just more likely to become physically and emotionally overstimulated.

However, HSPs also have higher levels of creativity, richer personal relationships, greater appreciation for beauty, and a rich inner life. Because behavior change involves learning, and learning involves the senses, an HSP may experience more emotional extremes when trying to change or learn a new behavior. This is normal. Feeling strong emotions in times of change is natural and does not indicate failure. While most people are not HSPs, everyone does fall somewhere on the sensory continuum. Due to the importance of our senses in learning new behaviors, we need to be aware of how (in)sensitive we are to sensory information.

What You Can Change (Nurture)

Although the ability to change your behavior is the result of nature and nurture and it is hard to explain one without the other, it will be more beneficial for you to focus on nurture because your genetic makeup is predetermined and hard (if not impossible) to change. Figuring out your genetic makeup, or biological traits, may be interesting and provide context to your situation, but that information is not enough to improve your mental health and cognitive function.

Instead, you need to focus on what you can control. Since much of your behavior and interaction with your environment is under your control, the decisions you make will modify the genes that get activated, which over time can improve your mental health. As a result, the remainder of Part 1 will focus on what you can change. There are two major behaviors under your control—your thought patterns and your social connection habits. While these are not the only areas you can change, they have an outsized impact on your ability to learn and adapt to new environments.

Thought patterns

While it may sound counterintuitive, the first area you can control is how you think. This is because of the interconnected brain-mind relationship; a change in one will trigger a change in the other. In other words, your brain's inner workings will reflect what you think and what you think will reflect your brain's inner workings. Your thought patterns are then the product of your brain's architecture and your mind's engagement. The way your brain is wired together and the thoughts in your mind can't be separated.

Historically, scientists thought that once you became an adult, the brain stopped growing and you were stuck with whatever architecture had been built. But over the last several decades, neuroscience research now supports the idea of neuroplasticity, or the brain's capacity to grow and change in response to life experiences. So instead of being permanently fixed in place, the brain's architecture is like soft plastic whose shape can be changed over time.

The neurons in your brain are like muscles in your body; they strengthen the more you use them and weaken when you don't use them. By thinking certain thoughts or using knowledge stored in long-term memory, you are reinforcing those connections, making them stronger. When you learn something new, you build new connections between neurons, modifying your brain's soft plastic architecture.

Learning something new is hard at first, but just like continuing to walk the same path through the woods will eventually create a new trail, a new behavior gets easier over time because you reinforce the new neural pathways. Similarly, the less you do something, the weaker the connections become over time. That is why it is hard to break a habit at first but becomes easier to avoid the old habit the less you do it. The connections between the bad habit's neurons deteriorate.

The most powerful combination is when you replace a bad habit with a good habit, like replacing scrolling through social media feeds with going for a walk outside. First, you will crave your old routine of grabbing your phone. But the longer you stick with the new routine of walking outside, the neurons associated with walking will strengthen while they simultaneously weaken for checking social media feeds.

> Although it takes a lot of time and effort, you can change your thought patterns through changing your behaviors.

Thought patterns are important because the amount and type of mental activity reflects your overall mental wellness efforts. We'll examine mental activity using anxiety and depression. These are the most common mental disorders in the United States, affecting millions of people each year. In fact, they are so common that people refer to them without knowing their definitions. Firefighters are at a higher risk than the public for developing both because of their regular exposure to trauma.[10] As a result, it is important that all firefighters understand these two conditions.

Anxiety

Anxiety is an emotion—a feeling of unease that occurs simultaneously with worried thoughts. As anxiety is part of our survival system, it helps to heighten our awareness in times of uncertainty like when walking alone at night. This is the fight-or-flight response. In these instances, feelings of anxiety are natural and welcomed. Anxiety, however, is supposed to be a short-term response to a dangerous situation. It moves to psychological distress when the anxiety becomes chronic. When an individual can no longer effectively cope with the symptoms, anxiety becomes a clinical psychological disorder.

When anxiety exists outside of an immediate threat, it is the result of an uncontrolled thought process. It is your brain in overdrive. Most people with anxiety disorders have a hyperactive fear network.[11] The fear network consists of three parts of your brain that regulate emotions, particularly negative emotions—the amygdala, insula, and cingulate cortex. While it isn't necessary to know the specific details of each part, it is important to understand that the fear network is part of the larger limbic system. Located in the center of your brain, the limbic system is responsible for regulating emotions and behavior. It is the emotional hub of the brain.

The increased brain activity from anxiety floods your body with cortisol, a stress hormone that activates the fight-or-flight response. Energy resources shift to help you survive a harmful or threatening situation like escaping from a burning building. For example, blood flow redirects from internal organs to muscles used for movement (e.g., arms, legs, etc.). The brain also pumps adrenaline into the bloodstream, increasing heart rate, breathing rate, and blood pressure. While helpful in times of need, an overactive fear network keeps you in a chronic state of stress, which over time makes you more susceptible to chronic diseases, including heart disease and cancer.

Depression

Depression, on the other hand, literally means a lowering of mood. Like anxiety, feelings of depression are part of a natural process for survival. If a threat is so bad or goes on for too long and there is no way to engage in fight or flight, the only option left is to freeze or immobilize. This disconnects us as a defense mechanism to dull the pain and emotions. It also has a metabolic effect, slowing the metabolism and changing how the body burns calories. Just like anxiety, this is only supposed to be a short-term effect. But when depression becomes chronic, people can have decreased self-esteem, increased self-criticism, and feelings of hopelessness and helplessness. It moves from psychological distress to a clinical mood disorder when feelings of sadness and loss of interest become persistent, disrupting a person's ability to function. Interestingly, elevated stress is both a cause and symptom of anxiety and depression.

Depression is on the other end of the thought pattern spectrum from anxiety; it causes brain activity to plummet, particularly in the limbic system and prefrontal cortex (the prefrontal cortex is responsible for all cognitive functions—it influences emotions through connections with the limbic system). Depression is a natural defense mechanism to bad situations that we cannot escape.[12] Also known as the faint-or-freeze response, the dorsal vagus nerve causes immobilization. As a result, the brain slows down and those in a depressed state feel detached, with their thoughts and emotions dulled.[13] Among all mental disorders, depression is one of the most potent in elevating an individual's risk for suicide.

While anxiety and depression represent opposite ends of the thought pattern spectrum, they can occur together, creating a unique combination of consequences.[14] The two conditions increase the size of the amygdala, making you even more sensitive to fear-based stimuli, always feeling there is a problem or something is wrong. But the problems are harder to solve because depression slows the prefrontal cortex. Eventually this can cause apathy about the situation, making it easy to lose hope. Since anxiety can increase the risk for suicide independent of depression, their combination creates an even higher risk of suicidal tendencies.[15]

Although most people are quick to self-diagnose, anxiety and depression are complicated and nuanced. They vary in severity from minor symptoms that are easy to overcome to debilitating disorders that require medical attention. Because anxiety and depression involve the mind and brain, neither condition can be diagnosed solely on biological symptoms like you can with a sprained ankle or the flu. As a result, diagnosis is partly subjective because it relies on an individual's self-reported symptoms and thought patterns. If all this was not complex enough, there is no one cause of anxiety or depression. Biology can be as much a factor as your interaction with the environment, making it difficult to generalize what relieves or prevents symptoms.

Social connection—individuals and groups

The second area you can control is your social connectedness with other people. As human beings, we have a fundamental need to belong.[16] It is natural to associate with other people and form relationships with them. In other words, we want to be in a community. Because we are social creatures, connecting with others is critical to our overall psychological well-being.

But we also need time by ourselves. This includes more than physical isolation like sitting in a room by yourself or taking a walk without anyone else. Real solitude is getting away from the input from other minds.[17] This means abstaining from anything produced by another person like listening to podcasts, reading books, or watching TV. While this may sound extreme, you need time for your mind to be alone with just your mind. Solitude isn't dependent on the environment. You can be alone with your thoughts as you hike in the woods or walk down a busy city street. This time alone encourages reflection, introspection, creativity, innovation, and self-discovery.

> Just like connecting with friends and family, regular doses of solitude will improve your psychological well-being.

The time you spend with others and in solitude are under your control. You must, however, be intentional with your social connection. It's easy to be around other people without connecting with them—you can be lost in your own thoughts or distracted by your phone. Similarly, you can be away from the influence of other people's minds but not use the time alone for thoughtful reflection. The more you can stay present with others and yourself, the more benefits you're going to reap.

Social connection—quality of engagement

To be socially connected with other people, you need to be able to build strong relationships. Empathy is the foundation for building a strong relationship. It allows you to understand, and vicariously experience, what another person is thinking or feeling. This mutual understanding allows for other important relationship characteristics to develop like trust, honesty, and love. Empathy is rare among living creatures—only a handful of mammals are known to be able to sense another's internal state. Neuroscientists believe that we are empathetic because of limbic resonance and regulation.[18]

As detailed in the next section, the limbic system is the part of your brain that controls emotion, behavior, and long-term memory. When we interact with another person, we sense the other person's internal state via mirror neurons. The two individual limbic systems connect and resonate together. As a result, you naturally adjust your own internal state to match the other person. This is why a person's attitude can be contagious. A happy person can lift up an otherwise grumpy group of people. Likewise, a stressed person can make everyone else around them stressed. Eventually, if you spend enough time around a person, your limbic systems can resonate with each other, or in other words, they sync together. For example, after living with a spouse for several years, a quick glance can tell you exactly what that person is feeling. This also happens to firehouse crews who have worked together for many years.

Although a lot of the processes underlying empathy are unconscious, as you'll see in this chapter, limbic resonance and regulation can be modified through practice. Although the ability to change your behavior is the result of nature and nurture and it is hard to explain one without the other, it will be more beneficial for you to focus on nurture because your genetic makeup is predetermined and hard (if not impossible) to change. Figuring out your genetic makeup, or biological traits, may be interesting and provide context to your situation, but that information is not enough to improve your mental health and cognitive function. Figure 4–3 provides more detail on the spectrum of limbic resonance and how it relates to firefighting. It is important to know that learning to be more empathic is critical to building strong, healthy relationships with family, friends, and coworkers.

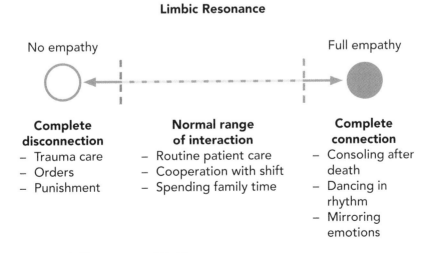

FIGURE 4–3. The spectrum of limbic resonance

Part 2: Deficiency and Excess

Thought Patterns

Thinking activates your brain. In a fraction of a second, electrochemical impulses travel up and down an interconnecting, crisscrossing web of neurons. If you're alive, the brain is active. Even in deep sleep, your neurons are sending and receiving messages. There is no off switch.

Although the brain never powers down, its activity fluctuates on a spectrum. Imagine your brain as a car. It can idle in a driveway doing the bare minimum to stay conscious, drive leisurely through a neighborhood lost in daydreams, or speed recklessly down a highway blinded by rage. From genetics to the environment, many things influence your thought patterns occurring at any amount.

Optimizing your brain activity requires an equilibrium between the extremes of your thought patterns. This starts with finding the right balance of doing and being. Doing represents purposeful, goal-orientated activities like remodeling your kitchen or running a marathon. Regardless of the activity, doing has structure—it is the means to an end. But it also builds competence, enhancing your feelings of self-worth. Alternatively, being activities are unstructured and contemplative. Being includes reflection, meditation, self-(re)discovery, and savoring the moment. Being is not a means to an end—it's listening to music or going on a walk. Regardless of the activity, being provides space to reflect on life and things bigger than yourself (fig. 4–4).

Doing	Being
Physical action is primary, reflection is secondary	Reflection is primary, physical action is secondary
– Sports – Eating – Cleaning	– Observing – Resting – Processing

FIGURE 4–4. States of doing and being

There are three strategies that can help you build "doing" and "being" activities into your life, which will optimize your thought patterns and, over time, improve your mental health. Those strategies are exercise, mindfulness, and mental rehearsal.

Exercise
Exercise is an excellent "doing" activity. As discussed in the physical fitness chapter, it has structure and works toward a specific purpose. Within your brain, exercise reduces and even prevents anxiety and depression. Research shows that a single workout decreases subclinical anxiety.[19] Going for a run or lifting weights after a stressful day will calm your body and mind. In addition, a regular exercise routine can reduce trait anxiety, a stable personality disposition that is hard to change otherwise. Exercise also eases depression symptoms in a dose-like relationship, meaning the more consistently you exercise, the less likely you will feel depressed.[20]

Despite the large amount of research, the exact mechanisms for why exercise reduces anxiety and depression are inconclusive. Most researchers agree that the process is complex, involving mechanisms of the brain and mind. It is also unclear whether one type of exercise is better than another. While most research studies examined the benefits of aerobic activity like running and walking, there is increasing evidence that weightlifting and mindful exercising (e.g., yoga, tai chi) work just as well.[21]

Exercise does more than reduce problematic thought patterns; it also optimizes cognitive function. It improves a range of important mental abilities like learning, remembering, problem-solving, and decision-making. These cognitive improvements are due to a hormone secreted by the brain called brain-derived neurotropic factor (BDNF). When you exercise, your brain increases BDNF production.

> Like a tune-up for your car, BDNF improves the brain's ability to function.

Mindfulness
Decades of research shows that mindfulness improves physical and psychological health as well as resilience. Mindfulness rewires thought patterns away from the extremes, making it one of the most effective strategies for reducing stress and anxiety. Because it gets results and improves performance, *Fortune 500* organizations and militaries around the world now integrate it into their training and corporate culture. For example, Nike, Google, and Procter & Gamble provide quiet rooms, yoga classes, and guided meditation to help reduce stress and refocus their employees during stressful days.[22] The U.S. Army uses

mindfulness training to help soldiers optimize their performance before and during deployment. It also helps reduce pain and stress related to post-deployment and post-traumatic stress disorder (PTSD).[23] Many of the techniques taught are the same ones detailed in this section.

However, its popularity has created a lot of confusion and misinformation, so it's easy to brush it off as just another fad. To provide some clarity, Dr. Shauna Shapiro has a clear and concise definition: intentionally paying attention in a kind, open way. As a clinical psychologist and professor at Santa Clara University, Dr. Shapiro refined this definition over her 20 years of studying and practicing mindfulness.

While the definition may sound vague, it is built around three core ideas—attention, intention, and attitude. As described in more detail later, these core ideas show how mindfulness is a type of lived mediation—a user-friendly way of taking control of your mind so you can focus without getting distracted. It's simpler than you may think. In fact, it might be easier to think of mindfulness as a foundation and not the activity itself. The three core ideas underlying mindfulness provide a framework to transform almost any activity into a "being" activity. As a result, mindfulness can be enjoying a cup of coffee on a quiet morning without worrying about the day ahead of you or listening to the sounds of nature on a hike instead of thinking about what you will eat for dinner in two hours.

Attention

The root of mindfulness is paying attention to the present moment. This means taking control of your mind to direct your attention to where you want it to go. The problem is that most people's minds wander almost 50% of the time. Instead of being in the present moment, we all live in a fog of time travel, daydreaming about the future or dwelling on the past. Being mindful clears the fog by keeping the mind in the present moment, increasing awareness of what is going on around you. While you will never stop your mind from wandering, practicing mindfulness will rewire the neurons in your brain to reduce your amount of time traveling.

Intention

You need to have a reason to be present in the first place. Intention uncovers your reasons for staying in the present moment. It reflects your values—your internal compass pointing to the things that matter most to you. For example, if you value being a caring spouse, then your intention to be present could be to focus on the needs of your significant other. You can have more than one intention, as long as each one is important to you.

Attitude

The final core idea is attitude, or how you pay attention. According to Dr. Shapiro's definition, you should pay attention in "a kind, open way." Mindfulness is not about observing life from a cold, detached third-person point of view. Instead, it is about being welcoming, inquisitive, and kind. This means that you observe experiences or thoughts (good, bad, or neutral) with an awareness that is nonjudgmental and inherently curious. Without kindness, mindfulness can become another way of being critical of yourself.

***

> Mindfulness is focusing your attention on the present moment with curiosity and nonjudgment. You do this for specific reasons that align with things you value. You then use mindfulness to engage fully in an activity.

Mindfulness is a great "being" activity, but it takes a lot of practice. In a world of distraction and instant gratification, learning how to just be with your own thoughts is not easy to do. But being mindful is different from meditating quietly for hours in a dark room. While formal meditation can help you become more mindful, mindfulness can be much more practical. Since mindfulness provides a framework, you can be mindful doing any activity no matter how mundane. Whether you're eating lunch or taking a walk, being present and not distracted will increase your psychological well-being.

Being more mindful involves slowing down. When you are rushing through the day, it's easy to be mindless, going through the motions without thinking. This can be especially hard at the fire station when things are busy. To start training yourself in mindfulness, it may require you to start on your off-duty days. But mindfulness is not just passive relaxation or observation of the present moment. Instead, it's intentional and active. You must choose to be present and pay close attention to what is happening around you without judgment or harsh criticism. Often, people jump to conclusions, thinking they know something without taking time to observe or question. This can lead to mistakes, arguments, and poor decisions. Over time, this type of thinking increases your stress response. If left unchecked, it can turn into chronic stress, which can quickly cascade into physical disease and psychological distress.

Being mindful does the opposite—it turns off the stress response, calming the body and mind. It increases psychological well-being. Strive to approach every situation and conversation with a present and open mind, suspending your initial conclusions and snap judgments.

Mental rehearsal

Your mind is powerful—it can visualize scenarios in rich detail. This allows you to create an experience in your mind and mentally rehearse it using all your senses. Many researchers refer to mental rehearsal as imagery because it uses your imagination. From daydreaming to visualizing a game plan, your imagination has the power to transform you to another place, or even into another person.

Mental rehearsal is a hybrid of "doing" and "being" because it combines elements of both. In fact, mental rehearsal's power lies in its flexibility. It allows you to plan for future events, reflect on past experiences, and even see things from another person's perspective. Since mental rehearsal relies only on your own mind, you can think about anything, anywhere, anytime.

> For firefighters, mental rehearsal turns the imagination into an important training tool that regulates brain activity by simultaneously decreasing stress and increasing self-confidence. This is important because managing stress can decrease or even eliminate anxiety.

Stress differs from anxiety. While they share many of the same physical and psychological symptoms, anxiety is excessive worrying that continues even when there are no stressors. As previously mentioned, stress can cause anxiety. For example, a recruit in the academy may be anxious about taking a test the next day because he does not feel prepared for it. The stress caused by not being prepared causes anxiety, but the anxiety itself is not stress. Stress isn't problematic per se; in fact, you need stress to grow and develop as a person. However, it becomes a problem when stress is chronic, with continually high levels over long periods of time. Mental rehearsal helps reduce stress and anxiety by providing a mechanism to plan, reflect, and see yourself being successful. As a result, it helps you deal with adversity by preparing for the unknowns.

There are individual differences when it comes to mental rehearsal. We all use our imaginations in different ways, so there is not a set of absolute rules for how to properly mentally rehearse or reflect. Instead, these three guidelines can help to sharpen your imagination:

1. To decrease anxiety and increase self-confidence about a situation, you need to imagine yourself doing the task with as much detail as possible. The more realistic the picture is in your mind, the more you improve your self-confidence for the task. For most, mental rehearsal is visual. But to tap into your imagination's full potential, involve all the senses—sight, sound, taste, touch, and smell.
2. Imagine failing as well as succeeding, because failure is going to happen. To prepare yourself, do not only visualize future successes. Imagine yourself struggling to complete the task. Additionally, reflect on actual experiences that did not go well, then think about how things could have been different. Mental rehearsal can also help you sharpen your own ideas. You can practice debating with yourself and looking for ways that your ideas might be wrong so that you can improve them. It is important not to end your mental rehearsals with failure. Instead, see yourself overcoming failures, difficulties, and obstacles.
3. Group input can enhance the effectiveness of mental rehearsal. An overlooked type of mental rehearsal is the After Action Review (AAR). Originally developed for the United States Army, the AAR requires both self and group reflection. During an AAR, individuals describe a previous experience, evaluate why it happened, and develop strategies to improve future performance. Although not relying as much on the imagination, AARs allow individuals to collectively review, discuss, learn, and plan from transpired events. This helps to reduce anxiety and increase confidence in the future.

> The National Fallen Firefighters Foundation explains the AAR in the 13th of 16 Firefighter Life Safety Initiatives (https://www.everyonegoeshome.com/16-initiatives/13-psychological-support/action-review/). NFFF offers free AAR training online (https://www.everyonegoeshome.com/training/action-review-aar/). NFFF also hosts an online training portal that offers free courses (www.fireherolearningnetwork.com).

Social Connection—Individual and Group

It is important to remember that mental health is not just the product of thought patterns, emotions, or genetic predispositions. According to Principle 3 in Kandel's framework, social, developmental, and environmental factors alter gene expression. In other words, there are factors outside of your brain's biology and your thought patterns that influence the development of your brain and mind. In fact, to achieve optimal levels of psychological well-being, you must also find a balance between two opposing social needs—solitude and relationships. Everyone needs meaningful relationships with other people, but also time alone away from other people (fig. 4–5). A deficiency in either will compromise your physical and mental health.

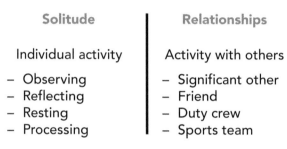

FIGURE 4–5. States of solitude and relationships

Solitude

At its core, solitude is being alone, away from the influence of another mind. We all have a need for solitude. Everyone needs time alone, regardless of how extroverted or socially inclined. If we prioritize relationships at the expense of solitude, we won't satisfy this need and our mental health will suffer.

Solitude is important because it provides social shelter.[24] It gives you space to focus the mind inward, allowing you to reenergize, reflect, process thoughts and emotions, plan for the future, and cope with disappointments. In this way, solitude acts as a relief valve for social pressures, easing stress that naturally builds up from being around other people.

Without having this needed space, your mind maintains an external focus, preoccupied with the people and experiences around you. This is dangerous because it's neglectful. When you work out constantly without giving your body the rest it needs, you don't reap the benefits of exercise because the body has no time to recover. If this is kept up for too long, you'll end up with an injury. Similarly, to reap the physical and mental benefits of relationships, you need to balance them with regular doses of solitude. Failing to do so will negatively impact your mental health.

Everyone needs solitude, but too much or too little will erode its positive benefits. Unfortunately, no universal number of hours exists for how long you should seek solitude each day, nor is there a Dunbar's number for solitude like there is for relationships (see the section on group relationships for more information). The amount of solitude you need is unique, based on your personality and life circumstances. It may take some experimenting to figure out how much you need.

Finally, do not confuse solitude with isolation. Both involve being alone, but they are opposite in every other way.[25] Solitude is a way to reenergize from your social relationships, while isolation is forced on you because of a lack of social relationships. As a result, solitude is voluntarily removing yourself from the influence of other minds for short periods of time; it's temporary. Isolation, however, emotionally drains you because you are involuntarily cut off from other people for an indefinite amount of time. Further, isolation can lead to a variety of physical and mental ailments.

> Solitude is being alone, while isolation is being lonely.[26]

Relationships

Just like with solitude, meaningful and close relationships play a vital role in human flourishing. We all need high-quality relationships. These include a network of close friends and family members, people who we care about and who care about us, and who we can lean on or learn from in challenging times. These relationships take time and energy to establish but will strengthen your mental health by providing feelings of mutual trust and obligation.

Individual relationships

Being in a healthy relationship with another person, whether it's romantic or platonic, has physical and mental benefits. These types of relationships include touch (e.g., hugging, holding hands, high-fiving) and promote feelings of warmth (e.g., trust, respect, care), both of which release oxytocin, a feel-good hormone that reduces stress by lowering blood pressure and cortisol levels. Oxytocin also promotes growth and healing throughout the body.[27]

To examine the health benefits of individual relationships in more detail, researchers studied married couples. They have found that people in a committed, loving marriage are less likely to die from cardiovascular disease, have better survival rates from cancer, and are at a lower risk of depression and Alzheimer's disease.[28] Although not known with as much detail, one could assume that close friendships and other types of committed relationships would have similar health outcomes.

Group relationships
Since high-quality relationships do take time and effort to create and maintain, not all social interactions will lead to them. There are several ways you can interact with people that will not meet your needs:

- Spending a little time with a lot of different people will be less satisfying than spending the same amount of time with less people.
- Being around a lot of people but not engaging with any of them will leave you feeling isolated.
- Knowing someone but not spending time with them, like the friend who calls once a year on your birthday, will feel superficial.

All these types of interactions will leave you unfulfilled because they do not lead to high-quality relationships.

This doesn't mean you should try to make as many high-quality relationships as possible. According to the psychologist Robin Dunbar, there is a limit to the number of high-quality relationships we can manage at one time.[29] Although the average size of a person's entire social network is approximately 150 people, we don't invest equally in every person. Instead, we devote about 40% of our total social effort to just 5 people (sometimes referred to as "Dunbar's number"). We spend another 20% of our effort on the 10 next most important people. As a result, approximately 60% of our social effort is divided between just 15 people. The low number of important people is reflective of the time and cognitive demands needed for maintaining high-quality relationships.

Focusing on our close network of high-quality relationships, we can experience positive physical and mental benefits. However, due to the significant effort and resources it takes to support high-quality relationships, having more interaction with the same people doesn't necessarily mean better health. Researchers found that increasing social interactions from weekly to daily did not correlate with better health, and for some, led to a higher risk for mortality.[30]

> Trying to spend too much time socializing creates more stress than it alleviates, lowering the quality of your relationship. Just like exercise, you can have too much of a good thing.

As a result, building—but not forcing—relationships with a small but trusted group of friends or family maximizes the health benefits of social interaction.

Doing so will ensure you create strong connections and share common interests, as opposed to social media connections that are weak and provide little help or encouragement. Evidence of this is seen among people in religious communities, who often have better health outcomes than nonreligious people. Researchers believe this is because religious affiliation puts an individual in a community with people who share similar beliefs, care about each other, and spend high-quality time with each other.[31] Nonreligious people can perhaps enjoy the same benefits by embedding themselves within a like-minded, supportive, community.

Optimizing individual and group relationships

Just like with brain activity, improving your psychological well-being is about finding balance between the extremes of solitude and relationships. The following three strategies will help you build relationships and solitude into your life, improving your overall mental health.

Strengthen connections

Relationships do not just happen. Each one takes time and effort, which is why you can only manage a few at a time. In addition, your relationships are like your muscles—you either use them or lose them. If you stop spending time with someone, your relationship with that person will weaken. Similarly, increasing the time you spend with someone will strengthen the relationship. While there are several factors that can determine the strength of your connections, two important factors are commonalties and physical distance.

Connecting with other people is like connecting fire hoses together. It's possible to connect a 5" diameter hose to a ½" diameter hose if you have the right connections, but the more adapters it takes to connect the two hoses, the weaker the overall connection. Your strongest connections are with people with whom you share the most in common. These similarities allow you to trust the other person and feel that they can understand you. This does not mean you shouldn't connect with people who are dissimilar; in fact, the different perspectives they bring may help make you a better person. Rather, it means that when people are different from you, it will take more time to accumulate shared experiences that eventually lead to high-quality relationships.

Physical distance also affects the connection between two people. Spending time with someone in person is the most powerful catalyst for building high-quality relationships. This is partly due to your mirror neurons. Your brain automatically mirrors the other person's emotions, which deepens your connection (you'll learn more about the power of mirror neurons in the section on empathy). In addition, you're more likely to share experiences in person. Common experiences help build commonalities with other people, which strengthen your connections. These experiences do not always have to be positive. In fact, hardship is a powerful bond. A platoon fighting in war or firefighters responding to a serious fire will share something that is indescribable to anyone else, creating a unique bond that is often unbreakable.

Given this information, it is not surprising that social connections weaken the farther removed you get from being in person. Seeing a friend or family member on video in real time is a good substitute, but not being in the same physical space limits the power of the shared experience. Similarly, talking on the phone is better than writing a letter, which is better than writing an email or text message. Cultivating high-quality relationships takes intention and effort. The more time you can spend with someone in person, the more experiences you build with them, the faster that relationship will strengthen and turn into a high-quality relationship.

Be intentional with technology
Although technology has made communication effortless with people across the world, it paradoxically has also caused people to interact less with each other. You might notice the effect of technology at the firehouse dinner table. While your shift might be in the same room and even around the same table, little interaction occurs because all of their attention is focused on their individual smart phones. As noted in the previous section, you can be around people but still be lonely because you are not interacting. Technology makes it easy to be entertained; but, if you are not careful, it will come at the expense of your high-quality relationships.

It is not just your social relationships that suffer from too much technology; it also negatively affects solitude. Technology may promise to satisfy all your needs, but too much of it will leave you feeling empty. The reason for this lack of satisfaction stems from the way you use technology. Instead of using it to free up more time for solitude, it becomes a means of distracting yourself when you feel bored. If you have ever spent an hour mindlessly scrolling through your phone, you know that this may feel good in the short-term, but eventually the distractor leads to frustration because it never fills the need to be alone. Even if what we are doing is educational, like reading a book or listening to a podcast, it prevents us from spending time alone away from other minds.

Technology has the unique ability to undermine both relationships and solitude—but it's not the problem. Technology isn't inherently evil. In fact, if used intentionally, smart phones and computers can add a lot of value to your life. Most people, however, fall into technology's trap because they don't think about why they are using certain apps or how a piece of technology will add something to their lives. This leads to technology creeping into every part of life, including relationships. It can replace high-quality relationships with shallow virtual connections and tempts you to trade moments of solitude for watching the hottest trending video.

> Technology can prevent you from getting the social shelter you need to reflect, organize, and synthesize your thoughts and get new insights and inspiration. Most firefighters are in a unique role; you have a lot of time off compared to other people. Don't fill all that extra time with entertainment, other people, or additional responsibilities like overtime or part-time jobs. Take advantage of the additional time to build in moments of solitude.[32]

To maintain high-quality relationships and enjoy moments of solitude, you must fight back against technology creep. Follow Cal Newport's philosophy of digital minimalism to determine what role each piece of technology plays in your life.[33] For each piece of technology, whether it is hardware or software, determine if it helps you achieve something you value.

- If it does provide value, then figure out how to maximize the value while minimizing the technology's ability to distract and waste your time.
- If it does not add value, then get rid of the technology, recycle the hardware, or delete the software.

As an example, let's say you value fitness and peer support. A fitness app on your phone can give you good workouts and connect you to a supportive online community. You can use that app for your workout each day and record your results, then spend a few minutes messaging your friends within the online community. Afterwards, close the app and don't look at it until your next workout. In this way, technology fulfills a need and provides value without taking away from your high-quality (in-person) relationships or solitude.

Develop helpful routines and habits
A routine is a sequence of actions that you follow on a regular basis. A common morning routine may involve exercising, showering, and eating breakfast. Similarly, a habit is a routine that is so engrained it is difficult to give up or modify. Habits are automatic—you're unlikely to deviate from them without a concerted effort.

Professionals use routines and habits to increase their productivity or efficiency. However, the benefits of routines and habits extend far beyond improving your ability to work. They are also an effective strategy to strengthen your relationships. At first, it may feel unnatural or robotic to develop routines and habits to build relationships. Most people govern their friendships through spontaneous interactions—for example, calling a friend after you hear a song that reminds you of them. Few people set a reoccurring reminder to call a friend on Saturday morning. It seems forced and unnecessary. The problem with this perspective is that the serendipitous moment may never come, whereas the scheduled call is sure to happen. Just like scheduled safety checks ensure the equipment works before responding to a fire, regularly scheduled connections ensure that you make it a habit to develop and maintain your social network.

Routines and rituals are just as important for cultivating solitude. In a world full of distractions, it is easy to never get a moment to yourself. By planning blocks of solitude into your day, even if it is only a few minutes, you will give yourself the space needed to reenergize. You do not need equal time devoted to relationships and solitude—find the right balance between the two extremes that works for you and commit to it. While it may seem hard in the beginning to find the time in your schedule, you will reap the benefits and quickly turn these new routines into lifelong habits.

Social Connection—Quality of Engagement

Relationships will not form without a genuine connection with another person. Empathy is the relationship's glue. It's the bond between two people caused by the ability to understand another person's feelings and experiences.

While empathy is necessary to form high-quality relationships, it doesn't help you in times of solitude. Being alone requires you to face your own thoughts and feelings. Some people try to avoid solitude because it can be uncomfortable and even painful at times. If you experience discomfort or pain during times of solitude, you may need to have more self-compassion. It's an inward-focused empathy that gives you the strength to process and accept whatever you feel, whenever you feel it. Don't confuse processing and accepting your feelings with acting on your feelings, which can get you into trouble if the feelings are not appropriate. This section will describe the importance of both empathy and self-compassion and how to improve them in your own life to enhance your mental and physical health.

Empathy

Empathy is the glue required to connect with people and build relationships. The Oxford Learner's Dictionary defines empathy as "the ability to understand another person's feelings, experiences, etc."[34] Empathy creates an emotional connection. It's a universal language, connecting you to close friends and strangers alike. Empathy is so critical to your mental health that a lack of it is characteristic of numerous psychological disorders.

Typically, empathy is innate. We don't have to learn or do anything special to be empathic. But the amount of empathy felt can vary among people and, depending on a variety of biological and environmental factors, some people may need to learn how to adjust their levels of empathy. For example, if you respond to trauma victims, you may need to reduce your level of empathy to do your job quickly and efficiently. If you are on your department's peer support team, you may need to increase your empathy when helping another firefighter get through an issue. As a firefighter, you will need to balance your empathy for your own mental health so that you can connect with other people but still do your job when things get tough.

Although complex, empathy can occur without consciously thinking about it. It's powerful. Just being close to another person triggers your empathic tendencies. Empathy is a primary reason you bond with the other firefighters on your shift, as well as an underlying motivation for wanting to help people who are in danger. Even though empathy is so common, most of the underlying biological mechanisms are still unknown. The complexity is the result of the mind and brain operating together, with both heavily influenced by the environment. But neuroscientists do have two theories of empathy—simulation and mind theory (fig. 4–6).[35]

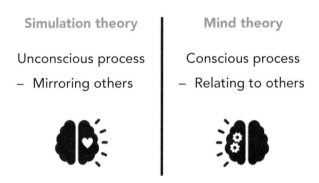

FIGURE 4–6. Engaging in simulation theory or mind theory

Simulation theory
Simulation theory suggests that empathy is an automatic process caused by mirror neurons.

- You recognize the emotions of another person and, unconsciously, your neurons produce the same emotion.

- Emphasis is on the brain's role in empathy—the neurons cause an emotional change that is then interpreted by your mind.

Mind theory
Mind theory posits that empathy is not an automatic process. Rather, it is a focused cognitive process.

- Empathy is your conscious decision to think about another person's thinking, and then being able to relate to it in some way.
- Emphasis is on the mind's role in empathy—thoughts cause your nervous system to produce an emotional change.

It may seem that these two theories are incompatible, but they interact and complement each other through a two-step process:

1. It starts with empathic concern. To immediately understand the emotions or experience of another person, there must be an automatic way that allows you to feel the emotions or jump into the experience as though it is happening to yourself. This process is simulation theory.
2. The process then moves to empathic accuracy. Once you can feel the emotions or imagine the experience, you can take a step back and evaluate the mental state of the other person. Reflecting on what is going on prevents you from reacting without first thinking about the situation. Taking time to think about the situation from different perspectives ensures that you understand what is going on. In addition, it gives you some space so you do not become so consumed by the situation that you are unable to make a rational decision. You want to be able to help the person, not just provide instant gratification or a quick fix to ease momentary discomfort.

Thus, the key to optimizing your empathy is not to pick between the simulation or mind theory. Rather, it is in your ability to find the middle ground, combining the theories to leverage the relationship between your brain and mind. This is important because if you only rely on your brain's automatic ability to mirror the emotions and experiences of other people, then you might overidentify with their emotions or experiences. This could lead to paralyzing pain or fear when treating a trauma patient on a call. In other words, you can mindlessly mirror the people around you without evaluating the usefulness of doing so. Instead, you should use your mind to process that information, evaluate it from multiple perspectives, and determine the best course of action.

Self-compassion
Empathy is not just important to your relationships; it is also critical to your self-acceptance, especially in times of reflective solitude. As firefighters, bad things will happen, especially in stressful situations. When you make a mistake or feel stressed, it's easy to either suck it up or wallow in self-pity. But these are extremes, neither of which improves your mental health, resilience, and overall performance. Instead, try choosing the middle ground by practicing self-compassion.

Negative emotions are common during stressful situations. Sometimes stress causes negative emotions like getting angry with yourself because you made a mistake during a hard time. Popularized by Dr. Kristin Neff, a professor of psychology, self-compassion is kindness shown to yourself when you suffer, fail, or feel inadequate. It is protection from harsh thoughts or emotions caused by stressful situations.[36]

Self-compassion is a specific type of mindfulness that goes beyond accepting experiences or thoughts without judgment. During a negative experience or thought, self-compassion encourages you to also embrace yourself with kindness. This may sound too touchy-feely, but neglecting self-compassion has serious consequences. Ignoring your problems by sucking it up never works—the pain will break free at some point. The longer you repress it, the worse the blowback. Self-compassion, however, allows you to acknowledge the pain and care for your emotional state. This allows you to take a step back to see that everyone goes through hard times. It is a natural part of life. By suspending judgment, you can see yourself in a more objective way, deepening your understanding and acceptance of who you are as a person. Importantly, self-acceptance opens you up to be more empathetic toward others.

> Self-compassion is about keeping things in perspective. Being your own worst critic will not be helpful if you cannot also forgive yourself, be optimistic about improving, and then try again.

Similarly, self-pity is just as bad as trying to ignore your problems because it is egocentric. Self-pity exaggerates your own suffering compared to others. Alternatively, self-compassion allows you to see that everyone struggles, possibly even with the same thing you do. Your experience will always be more like others than you think. At its core, self-compassion is empathic accuracy applied to yourself. By taking a broader perspective, self-compassion frees you to acknowledge that struggling is part of being human. By letting go of being perfect, you will improve your mental health.

Optimizing quality of engagement

Optimizing high-quality relationships and solitude requires empathy and self-compassion. As with the other skills described in this chapter, they are not easy to obtain and may not always be your default. There are three strategies that can help improve your empathy and self-compassion. Those strategies include changing your perspective, supportive self-talk, and SWOT (Strengths, Weaknesses, Opportunities, and Threats) analysis.

Change your perspective

To improve your empathy, you need to consider the same situation from multiple perspectives. While your brain automatically mirrors the emotions of another person, you maintain conscious control of your thoughts. Thus, when you begin to feel the emotions of another person, try to imagine what that person is also thinking. Trying to inhabit their internal world will give you a greater perspective on why they feel that emotion. But as simple as it sounds, in the moment, it can be hard to take a mental step back, especially if it is emotionally charged. In those instances, when you feel yourself becoming too engulfed in the emotion to see different perspectives, regain control by removing yourself from the situation.

You can also try changing your self-perspective. Typically, this involves reversing how you reflect on the past, particularly when you have failed. For most, the default response involves some self-criticism. This can be helpful if you focus the criticism on the mistake without personal assaults or harsh judgments. Unfortunately, this is rarely the case. You probably say things to yourself that don't help you learn from what went wrong. This type of reflection can also raise, rather than reduce, your stress.

Instead, imagine yourself through your best friend's perspective. What would that person say to you? How would they encourage you? A good friend is an advocate who helps you overcome the situation, not a critic who just points out your flaws without giving you a hand. Yet, many people do not treat themselves this way. Be aware of your perspective so you can shift from a self-critic to a self-advocate.

> "The mark of a sharp mind is the ability to entertain a competing thought, without necessarily accepting it." —Aristotle

Supportive self-talk
Paired with shifting your perspective, your inner dialogue must be supportive. What you say to yourself matters, so you need to optimize your self-talk. You may be your harshest critic, but you should also be your biggest supporter. Supportive self-talk isn't ignoring your mistakes or pretending they didn't happen. Rather, your self-talk needs to match your perspective as a self-advocate.

In addition, to optimize your moments of solitude, your inner dialogue needs to be kind and not overly critical. This is especially important when you struggle with a difficult task or fail at something important. Again, being kind isn't overlooking mistakes, and it especially doesn't mean pretending everything is fine when everything is falling apart. Instead, helpful self-talk provides neutral and supportive feedback. During the action, focus on cues that are relevant to the task without dwelling on the past or thinking about the future. After the action, regardless of the outcome, tell yourself what you learned and how you can improve. Celebrate victories and move on from failures.

Supportive self-talk should also put whatever you are trying to accomplish in perspective. For example, if you are struggling to learn a new skill, you're probably not as bad as you think. Make sure your self-talk isn't exaggerating the importance of the situation or the consequences for failure, especially in a training scenario. It may help to get another person to give you feedback. It is also important to remember that everyone struggles—no one is perfect. Failure is part of being human, so you are not alone.

This is also not an excuse for failure or not making progress on a goal. Rather, choosing to talk to yourself in a supportive manner and putting things into perspective is part of a growth mindset, seeing failures and mistakes as learning opportunities. It is impossible to learn from your mistakes if you are harsh and overly critical. To prevent this from happening, a good rule is to talk to yourself like you would to a good friend going through the same situation. Be supportive, provide unbiased feedback, and avoid inflating the magnitude of failure. In short, be supportive of yourself. This can include critiquing yourself, but there is a difference between constructive and destructive criticism.

SWOT analysis
Traditionally, a SWOT analysis is a framework used to evaluate a company's competitive position and develop strategic planning. It assesses internal and external factors, as well as current and future potential by prompting you to consider all sides of the story. Although it seems far-fetched, a SWOT analysis can help you develop empathy and self-compassion (fig. 4–7).

A natural survival instinct as a human is to look for the danger (i.e., your weaknesses or threats against you). Conducting a SWOT analysis identifies your weaknesses and threats, but also your strengths and opportunities.[37] While

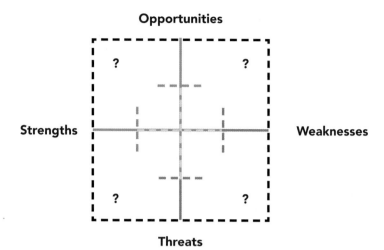

?: Your behavior changes based on how you perceive reality. It is natural to identify threats and weaknesses first for survival purposes. How much time do you spend considering strengths and opportunities?

FIGURE 4-7. SWOT analysis

this is helpful, a SWOT analysis is not about making lists for yourself under each letter of SWOT. Instead, it's a planning document used to consider courses of action based on how your strengths and weaknesses converge with opportunities and threats. For instance, if you identified that you have a short temper as a weakness, and someone you dislike was just transferred to your shift (a threat), you can plan a course of action to limit one-on-one interaction with that person. To help you make your own plan, here is an explanation and example for each SWOT factor:

- **Strengths**: What you excel at and what unique qualities separate you from other people. For example, you might be a good firehouse cook.
- **Weaknesses**: What stops you from performing at your optimum level. These are areas where you can improve. For example, you have a short temper.
- **Opportunities**: Favorable external factors that are advantageous to you. For example, cooking gives you an opportunity to spend time alone in the kitchen away from the person you dislike, and you control the meals, not the other guy.

- **Threats**: Factors that have the potential to harm you or your performance. For example, you may get heckled for your dinner by the new guy and you need to prepare to not lose your temper over that.

Although intended for organizations, applying a SWOT analysis on yourself can help improve your plans for managing relationships and having more constructive moments of solitude. It's a structured reflection and planning activity that helps you understand different aspects of yourself.

The better you know yourself, the better you can develop strong relationships, avoid negative ones, and plan courses of action for relationships, life experiences, and emergencies.

When conducting a SWOT analysis, or any type of reflection activity, avoid using digital devices like your smartphone, tablet, or computer. Instead, do the SWOT analysis by hand with pen and paper. Compared to entering the same information on an electronic device, researchers from the University of Tokyo discovered that writing by hand activates areas in the brain responsible for language, imagery, and memory.[38] The researchers believe using pen and paper helps you learn the information better, making it more easily retrieved from memory. In turn, this improves your ability to use and be creative with that information.

Living the balanced life

Controlling your thought patterns and efforts at social connection provides a foundation for your mental health. But how do they work together? Dr. Hadassah Littman-Ovadia created a model for a balanced adult life.[39] The model requires balancing the two pairs of opposing needs described earlier in this section—doing versus being and solitude versus relationships. Combining these two pairs creates four categories: solitary doing, communal doing, solitary being, and communal being (fig. 4–8).

- **Solitary Doing**: Engaging in activities alone. May be beneficial for completing complex and challenging tasks.
- **Communal Doing**: Engaging in activities with others. May be more enjoyable or pleasurable than solitary activities, especially if a group reaches a flow state. Best for highly interdependent activities.
- **Solitary Being**: Alone in a contemplative state. Can be achieved through mindfulness-based activities. Improves physical

and mental health, cognitive skills, self-discovery, and self-transformation.
- **Communal Being**: Shared contemplative experience. More than just occupying the same physical space, those involved shape each other's experience through actions like two or more people having a deep conversation. It increases intimacy with others, deepening relationships.

To optimize your mental wellness efforts, you will want to make an effort to spend some time in each category. The exact amount of time needed in each category will vary by individual. For example, some people may prefer slightly more solitary activities compared to communal activities. But too much time spent in any one category at the expense of the others will lead to an imbalance and dissatisfaction with life. When combined with other genetic and environmental factors, a severe imbalance can result in psychological distress, or cascade into a more serious psychological disorder. A balanced life leads to better mental health.

Although Dr. Littman-Ovadia didn't include it in her model, quality of engagement could be a third dimension to the model. You can engage in each of the four combinations with very little effort, moderate effort, or a lot of effort (this is portrayed in figure 4–8 by the size of the dot). The more effort you make, the better the results. For example, the more effort you put into being empathetic in communal settings, the more powerful the communal experience. Similarly, the more effort you put into self-compassion in solitary settings, the more restorative the experience. The key to a balanced life lies in quality of experiences, not just the quantity of them.

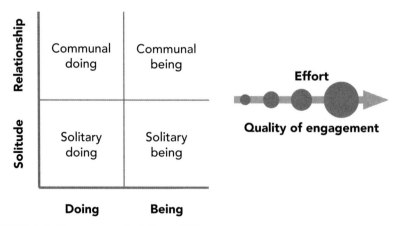

FIGURE 4–8. Components of a balanced life

Part 3: Micro and Macro

> NFPA 1500: Standard on Fire Department Occupational Safety, Health, and Wellness Program and NFPA 1583: Standard on Health-Related Fitness Programs for Fire Department Members provide standards on behavioral health, general wellness, managing stressful events, and member assistance programs. Information on those standards can be found in NFPA 1500 chapter 12 and 13, and in NFPA 1583 chapter 8 (Section 8.1.4). Additional information can be found in NFPA 1500 Annex A (Section A.12 and Section A.13).

Part 1 introduced Kandel's framework to describe the role of nature and nurture on mental health. Although you cannot change your genetic composition, you can improve your mental health by maximizing psychological well-being and minimizing psychological distress. Part 2 showed how this is done by finding the virtuous middle ground in your thought patterns and social connections.

This section will summarize how engaging in mental wellness can improve your quality of life and provide protection not just for your mental health, but your physical health as well. According to Kandel's framework, Principle 4 and 5 state that neurobiological changes in the brain occur by modifying your behavior. The decisions you make about your lifestyle play a major role in your mental health. A change in what you do or think each day rewires your neurons. If it's positive, you'll increase psychological well-being and cognitive functioning. If it's negative, you'll increase psychological distress.

Adopting the strategies described throughout the chapter will improve your mental health by optimizing your thought patterns and strengthening your connections with yourself, other individuals, and groups. These changes are not an end in themselves. Rather, as detailed later in the chapter, they positively affect your quality of life. The WHO defines quality of life as:

> *An individual's perception of their position in life in the context of the culture and value system in which they live and in relation to their goals, expectations, standards and concerns. It is a broad ranging concept affected in a complex way by the person's physical health, psychological state, personal*

> *beliefs, social relationships and their relationship to salient features of their environment.*

Quality of life is broad and complex, encompassing many of the concepts discussed throughout this book. It's also subjective, based on your perception and overall evaluation of life. As a result, quality of life can differ significantly between two people experiencing similar situations. But regardless of the individual or situation, improving your mental wellness using the strategies discussed in this chapter will improve your quality of life because you are pursuing behaviors in the virtuous middle instead of the extremes. Engaging in mental wellness can improve the way your brain works, and even help your brain to grow.

Cognitive Functioning

Optimizing your thought patterns and social relationships will improve your mental abilities. Some examples include learning, thinking, reasoning, remembering, problem-solving, decision-making, imagination, perception, and attention. These are essential skills firefighters need when responding to any emergency.[40]

One of the biological reasons for improvement is because of a less active amygdala. As you optimize your thought patterns, your fear network becomes less active because of lower levels of stress and anxiety. This allows you to feel more relaxed and less rushed in emergency situations, giving you more time to evaluate information. At the same time, as activity increases in the prefrontal cortex, your ability to think, remember, and make accurate judgments will improve. These two changes lead to better decisions, helping you perform your job better, which as a firefighter means keeping more people and property safe.

Neurogenesis and Neuroplasticity

The brain is not static. Neurogenesis and neuroplasticity are two of the mechanisms that change the brain over time. Neurogenesis is the ability to create new neurons (when needed) and replace old ones. Just like muscles in the body, the brain can become bigger or smaller depending on how it's used. For example, as noted earlier in the chapter, depression shrinks the size of the prefrontal cortex. This decreases cognitive function, leading to a host of negative outcomes like memory loss and the inability to cope with stress. But because of neurogenesis, when you increase your mental wellness by being mindful or developing high-quality relationships, the prefrontal cortex can be rebuilt.[41]

Not only can your brain make new neurons, but it can also make new connections between neurons. Neuroplasticity is the brain's ability to change and adapt because of experience, allowing you to learn new things, unlearn old things, improve existing cognitive abilities like memory or attention, and even recover from traumatic brain injuries.[42]

When you start one of the new behaviors described in Part 2, like supportive self-talk or engaging in high-quality relationships, your brain goes through neuroplasticity; it creates new connections.[43] Each time you practice supportive self-talk after any constructive self-criticism you might do, it strengthens the connection until it becomes engrained like a well-worn hiking trail through your brain. As you learn to quit any self-pity or destructive self-criticism, the brain prunes away the connections related to negative thinking, making it less likely you will do it in the future.

While neurogenesis and neuroplasticity have a lot of positive results, it is important to understand that there are limits to what they can do. The brain is not infinitely malleable, and some injuries cause irreversible damage. In addition, the brain can be influenced by psychoactive drugs or pathological disorders. These can have detrimental effects on the brain. However, as a firefighter, you should know that you can make a difference in your mental health through your efforts at changing your behavior.

Neurological Diseases

Improved mental wellness is associated with a lower risk of developing neurological diseases, specifically Alzheimer's and Parkinson's disease. Alzheimer's disease is a severe form of cognitive impairment, encompassing a wide range of cognitive functions like memory, attention, language, and visuospatial skills. Although much is still unknown, it appears that managing stress and depressive symptoms helps to prevent the cognitive declines associated with Alzheimer's.[44] Since optimizing your thought patterns and high-quality relationships do both of those things, they can have a protective effect against developing Alzheimer's in the future.

Parkinson's disease is a progressive disease of the nervous system in which the nerve cells gradually degenerate, causing a person to lose control of physical movement. Similar to Alzheimer's, scientists are still researching the causes and strategies of prevention for Parkinson's disease. However, some evidence suggests that mindfulness-based activities improve areas of the brain negatively affected by Parkinson's.[45]

The Big Three (Heart Attack, Cancer, and Suicide)

Psychological autopsy studies have found that 90% of suicide victims have psychological disorders like depression, anxiety, or addiction.[46] While these diseases may require medical attention that extends beyond the strategies described in Part 2, there is a connection between suicide, thought patterns, and mental wellness. For those not currently suffering from serious psychological disorders, optimizing your thought patterns and strengthening your connections with others by using the strategies described earlier in the chapter can help protect you from developing suicidal thoughts.

Social isolation and loneliness are important risk factors associated with suicidal thoughts and behaviors.[47] As a result, strengthening relationships may help prevent suicidal tendencies. While suicide is a complex topic that often extends beyond thought patterns and relationships, you are limited to what you can control, and how you think and how you interact are two of the biggest mental wellness behaviors within your control.

It is critical to remember that your brain and mind are interconnected, and that your brain is part of your physical body. If you experience repeated stress, especially when this stress is not related to survival and becomes chronic due to negative thought patterns or poor social connections, your psychological stress can eventually damage your body and pave the way for cancer and heart disease.

Researchers are continuously updating the research on the connections between psychological stress and cancer.[48] In large meta-analysis studies, it appears that stress is linked to the growth of cancer and cellular aging. The main reason seems to be that chronic stress increases the inflammatory response in your body while also lowering your immune surveillance, which then leads to the development and growth of tumors.[49] The more stress hormones that are circulating in your body, the worse your immune system functions.

A more well-established relationship already exists between stress and heart disease. Whether it is at work or at home, chronic stress greatly increases your odds of having a heart attack.[50] The INTERHEART study, which took data from 52 countries, found that chronic stress increases your chances of having a heart attack by over 200%![51] This powerful statistic demonstrates that your mind and brain have a powerful effect on your body and health outcomes.

Placebo Effect

Perhaps the most unexpected phenomenon of mental wellness is the placebo effect. The placebo effect is like the reverse of chronic stress on your body. It is the positive effect of your mind and brain on your body. It works through conditioning. In cause-and-effect relationships, you expect certain results based on previous experience. The more you've experienced a certain outcome, the more likely you'll expect that outcome again. For example, if taking an aspirin got rid of a headache every time you've taken it, your expectation of relief can cause you to actually feel relieved even if you are given a placebo pill.

But the placebo effect is even more powerful. Even without previous experience, if you expect a certain outcome, then it is more likely to happen. Researchers found evidence of this effect when reviewing the results of 19 randomized clinical trials testing the effectiveness of antidepressants. The placebo worked just as well as four of the six most popular antidepressants, and only in the most severe cases of depression did the prescription drugs work better.[52] Similarly, when testing a treatment for Parkinson's disease, researchers told people in the placebo group that the "drug" (saline solution) would significantly improve their symptoms. The placebo led to positive changes in the brain, not unlike taking a dose of a drug approved to treat Parkinson's disease.[53]

At its core, the placebo effect is positive thinking. The power of an expectation to impact the outcome shows the connection between the mind and brain. As the studies on the placebo effect demonstrate, thinking has a significant impact on the health of the body. Similarly, several other research studies have shown that optimists, or people with positive attitudes, were healthier, suffered from less pain, had more energy, and lived longer.[54]

***

All this evidence shows that the way you think matters. Thought can be a powerful medicine or a poison. Your thoughts and attitudes can prevent and treat neurological (e.g., Parkinson's), psychological (e.g., depression and anxiety), and even physical health issues (e.g., heart disease and cancer).[55] Adopting simple strategies like changing your perspective, supportive self-talk, self-compassion, and mental rehearsal can drastically change the way you think about situations.

Conclusion

1. Mental health is the result of your biology and your interactions with your environment.
2. You cannot change your intelligence or sensory function.
3. You can change your thought patterns and social connections.
4. Exercise and mindfulness-based wellness efforts improve your mental health and reduce the risk of developing anxiety and depression.
5. Balance your time between two opposing social needs—solitude and social relationships.

6. Optimizing your time alone and time with others requires empathy and self-compassion.
7. Your individual mental health and the individuals and groups you engage with have a reciprocal relationship.

Mental wellness is vital to your overall health and performance as a firefighter. Due to the nature of the job, you are exposed to trauma and stress that can negatively influence your personal and professional life. But it is not always easy to understand what to do about mental health because it is the product of complex biological, psychological, environmental, and behavioral processes. Fortunately, tangible steps exist that you can take to protect and improve your mental health by altering your thought patterns and social connections.

Taking control of your mental wellness will help you contribute to a stronger, more resilient fire service. But the responsibility does not rest on you alone. Your fire department must take active measures to educate you and your shift mates on how to optimize your mental wellness. Without the support from the fire chief and other officers, it will be difficult to overcome the stigma and barriers surrounding mental health in the fire service.

Questions

1. What is the definition of mental health?
2. What is the difference between psychological distress and psychological well-being?
3. What is the difference between the brain and the mind?
4. What are Dr. Kandel's 5 principles of the brain-mind relationship?
5. What are two things that you cannot change about your mental health?
6. What are two things you can change about your mental health?
7. What is limbic resonance?
8. What is the difference between doing and being?
9. What is Dr. Shapiro's definition of mindfulness?
10. What are the three core ideas of Dr. Shapiro's teaching on mindfulness?
11. What are the three guidelines to sharpen your imagination using mental rehearsal?
12. What are the three different strategies for optimizing social connection?

13. What is the difference between empathy and self-compassion?
14. What is the difference between simulation theory and mind theory?
15. What are the three different strategies for optimizing quality of engagement?
16. What is Dr. Littman-Ovadia's model of a balanced adult life?
17. What is neuroplasticity?
18. What are the three ways that balanced mental activity can protect your health?
19. What chapters in NFPA 1500 relate to behavioral health?
20. Which chapter in NFPA 1583 discusses the topic of member assistance programs?

Notes

1. World Health Organization, *Promoting Mental Health: Concepts, Emerging Evidence, Practice: Summary Report* (World Health Organization, 2004).
2. National Center for Chronic Disease Prevention and Health Promotion, Division of Population Health, "About Mental Health," Centers for Disease Control and Prevention, April 25, 2023, https://www.cdc.gov/mentalhealth/learn/index.htm.
3. Ping Li, Jennifer Legault, and Kaitlyn A. Litcofsky, "Neuroplasticity as a Function of Second Language Learning: Anatomical Changes in The Human Brain," *Cortex* 58 (2014): 301–24.
4. Eric R. Kandel, "A New Intellectual Framework for Psychiatry," *American Journal of Psychiatry* 155, no. 4 (1998): 457–69, https://doi.org/10.1176/ajp.155.4.457. Dr. Kandel published this article two years before winning the Nobel Prize in Physiology or Medicine for his work identifying the physiological changes that occur in the brain during the formation and storage of memories.
5. Kandel, "A New Intellectual Framework."
6. Jordan W. Smoller, "Psychiatric Genetics Begins to Find Its Footing," *American Journal of Psychiatry* 176, no. 8 (August 2019): 609–14, https://doi.org/10.1176/appi.ajp.2019.19060643; Emily Deans, "Genetics and Mental Illness: Understanding the Genetics of Psychiatry Is Closer Than Ever," *Psychology Today*, September 30, 2019, https://www.psychologytoday.com/us/blog/evolutionary-psychiatry/201909/genetics-and-mental-illness. For a simpler explanation of Smoller's article, read Dean's article "Genetics and Mental Illness" on *Psychology Today*.
7. Angela M. Brant et al., "The Nature and Nurture of High IQ: An Extended Sensitive Period for Intellectual Development," *Psychological Science* 24, no. 8 (2013): 1487–95, https://doi.org/10.1177/0956797612473119; Chrissie Long, "The Nature and Nurture of High IQ: An Extended Sensitive Period for Intellectual Development,"

The Journalist's Resource, December 17, 2013, https://journalistsresource.org/education/nature-nurture-intellectual-development-children/. For a brief summary of Brant et al.'s article, read Long's review at *The Journalist's Resource*.
8. Ladan Shams and Aaron R. Seitz, "Benefits of Multisensory Learning," *Trends in Cognitive Sciences* 12, no. 11 (2008): 411–17, https://doi.org/10.1016/j.tics.2008.07.006.
9. "Highly Sensitive Person," *Psychology Today*, accessed July 20, 2023, https://www.psychologytoday.com/us/basics/highly-sensitive-person. An estimated 15%–20% of the population are thought to be highly sensitive. For a simpler overview of a highly sensitive person and sensory processing sensitivity, read the article on *Psychology Today*.
10. Daniel J. Paulus et al., "The Unique and Interactive Effects of Anxiety Sensitivity and Emotion Dysregulation in Relation to Posttraumatic Stress, Depressive, and Anxiety Symptoms Among Trauma-Exposed Firefighters," *Comprehensive Psychiatry* 84 (2018): 54–61, https://doi.org/10.1016/j.comppsych.2018.03.012.
11. Kathrin Holzschneider and Christoph Mulert, "Neuroimaging in Anxiety Disorders," *Dialogues in Clinical Neuroscience* 13, no. 4 (2022): 453–61, https://doi.org/10.31887/dcns.2011.13.4/kholzschneider.
12. Stephen W. Porges, "The Polyvagal Theory: New Insights into Adaptive Reactions of the Autonomic Nervous System," *Cleveland Clinic Journal of Medicine* 76, no. 4 suppl. 2 (2009): S86–90, https://doi.org/10.3949/ccjm.76.s2.17; Alison Escalante, "We've Got Depression All Wrong. It's Trying to Save Us: New Theories Recognize Depression as Part of a Biological Survival Strategy," *Psychology Today*, December 22, 2020, https://www.psychologytoday.com/us/blog/shouldstorm/202012/we-ve-got-depression-all-wrong-it-s-trying-save-us. For a less scientific description of Porges's article, read Escalante's article on *Psychology Today*.
13. Paul B. Fitzgerald et al., "A Meta-Analytic Study of Changes in Brain Activation in Depression," *Human Brain Mapping* 29, no. 6 (2008): 683–95, https://doi.org/10.1002/hbm.20426; Daniela A. Espinoza Oyarce et al., "Volumetric Brain Differences in Clinical Depression in Association with Anxiety: A Systematic Review with Meta-Analysis," *Journal of Psychiatry and Neuroscience* 45, no. 6 (2020): 406–29, https://doi.org/10.1503/jpn.190156.
14. Anna S. Engels et al., "Co-Occurring Anxiety Influences Patterns of Brain Activity in Depression," *Cognitive, Affective, & Behavioral Neuroscience* 10, no. 1 (2010): 141–56, https://doi.org/10.3758/CABN.10.1.141.
15. Peter J. Norton, Samuel R. Temple, and Jeremy W. Pettit, "Suicidal Ideation and Anxiety Disorders: Elevated Risk or Artifact of Comorbid Depression?" *Journal of Behavior Therapy and Experimental Psychiatry* 39, no. 4 (2008): 515–25, https://doi.org/10.1016/j.jbtep.2007.10.010.
16. Roy F. Baumeister and Mark R. Leary, "The Need to Belong: Desire for Interpersonal Attachments as a Fundamental Human Motivation," *Psychological Bulletin* 117, no. 3 (1995): 497–529, https://doi.org/10.1037/0033-2909.117.3.497.

17. Raymond M. Kethledge and Michael S. Erwin, *Lead Yourself First: Inspiring Leadership Through Solitude* (Bloomsbury Publishing USA, 2017).
18. Paul D. MacLean, "Evolutionary Psychiatry and the Triune Brain," *Psychological Medicine* 15, no. 2 (1985): 219–21, http://www.doi.org/10.1017/S0033291700023485.
19. Steven J. Petruzzello et al., "A Meta-Analysis on the Anxiety-Reducing Effects of Acute and Chronic Exercise: Outcomes and Mechanisms," *Sports Medicine* 11, no. 3 (1991): 143–82, https://doi.org/10.2165/00007256-199111030-00002.
20. James A. Blumenthal, Patrick J. Smith, and Benson M. Hoffman, "Is Exercise a Viable Treatment for Depression?" *ACSM's Health & Fitness Journal* 16, no. 4 (Summer 2012): 14–21, https://doi.org/10.1249/01.FIT.0000416000.09526.eb.
21. Kaushadh Jayakody, Shalmini Gunadasa, and Christian Hosker, "Exercise For Anxiety Disorders: Systematic Review," *British Journal of Sports Medicine* 48, no. 3 (2014): 187–96, https://www.doi.org/10.1136/bjsports-2012-091287; Blumenthal, Smith, and Hoffman, "Is Exercise a Viable Treatment."
22. James Duffy, "10 Big Companies That Promote Employee Meditation," *More Than Accountants*, January 2, 2020, https://www.morethanaccountants.co.uk/10-big-companies-promote-employee-meditation/. Read for a summary of different corporate mindfulness and meditation programs.
23. "Mindfulness for the Military," *Human Performance Resources by CHAMP*, April 28, 2020, https://www.hprc-online.org/mental-fitness/sleep-stress/mindfulness-military. Read for more information on what the U.S. Army is doing with mindfulness training.
24. Dongning Ren, "Solitude Seeking: The Good, the Bad, and the Balance" (PhD diss., Purdue University, 2016).
25. Michael Schreiner, "Isolation Versus Solitude," *Evolution Counseling*, June 27, 2013, https://evolutioncounseling.com/isolation-versus-solitude/.
26. Julianne Holt-Lunstad et al., "Loneliness and Social Isolation as Risk Factors for Mortality: A Meta-Analytic Review," *Perspectives on Psychological Science* 10, no. 2 (2015): 227–37, https://doi.org/10.1177/1745691614568352; Gina Agarwal et al., "Social Factors in Frequent Callers: A Description of Isolation, Poverty and Quality of Life in Those Calling Emergency Medical Services Frequently," *BMC Public Health* 19 (2019): 1–8, https://doi.org/10.1186/s12889-019-6964-1.
27. Kerstin Uvnäs-Moberg and Maria Petersson, "Oxytocin, a Mediator of Anti-Stress, Well-Being, Social Interaction, Growth and Healing," *Zeitschrift für Psychosomatische Medizin und Psychotherapie* 51, no. 1 (2005): 57–80, https://doi.org/10.13109/zptm.2005.51.1.57.
28. "Marriage and Men's Health," *Harvard Health Publishing*, June 5, 2019, https://www.health.harvard.edu/mens-health/marriage-and-mens-health.
29. R.I.M. Dunbar, "The Anatomy of Friendship," *Trends in Cognitive Sciences* 22, no. 1 (2018): 32–51, https://doi.org/10.1016/j.tics.2017.10.004.
30. Olga Stavrova and Dongning Ren, "Is More Always Better? Examining the Nonlinear Association of Social Contact Frequency with Physical Health and Longevity," *Social Psychological and Personality Science* 12, no. 6 (2021):

1058–70, https://doi.org/10.1177/1948550620961589. The study found that increasing contact with friends, colleagues, and neighbors from never to monthly was associated with a 10% decrease in mortality risk. However, increasing contact from monthly to daily contact led to an 8% increase in mortality risk.
31. David G. Schlundt et al., "Religious Affiliation, Health Behaviors and Outcomes: Nashville REACH 2010," *American Journal of Health Behavior* 32, no. 6 (2008): 714–24, https://doi.org/10.5993/AJHB.32.6.15.
32. Thomas Oppong, "For a More Creative Brain, Take Breaks: Your Brain Needs Downtime to Remain Creative and Generate Its Most Innovative Ideas," *Inc.*, May 30, 2017, https://www.inc.com/thomas-oppong/for-a-more-creative-brain-take-breaks.html; Jackie Coleman and John Coleman, "The Upside of Downtime," *Harvard Business Review*, December 9, 2012, https://hbr.org/2012/12/the-upside-of-downtime; Srikant Chellappa, "The Importance of Downtime," *Engagedly*, November 12, 2019, https://engagedly.com/the-importance-of-downtime/. Read these articles for more information on the importance of rest and solitude.
33. Cal Newport, *Digital Minimalism: Choosing a Focused Life in a Noisy World* (Penguin Random House, 2019).
34. Oxford Learner's Dictionary, s.v. "empathy (n.)," accessed July 24, 2023, https://www.oxfordlearnersdictionaries.com/us/definition/english/empathy.
35. Lian T. Rameson and Matthew D. Lieberman, "Empathy: A Social Cognitive Neuroscience Approach," *Social and Personality Psychology Compass* 3, no. 1 (2009): 94–110, https://doi.org/10.1111/j.1751-9004.2008.00154.x.
36. Kristin Neff, *Self-Compassion: The Proven Power of Being Kind to Yourself* (William Morrow Paperbacks, 2011). You can find more information on self-compassion on Neff's website, https://self-compassion.org/.
37. If you are interested in doing a SWOT analysis on yourself, you can get more information and a template at https://www.ncbi.nlm.nih.gov/books/NBK537302/.
38. Keita Umejima et al., "Paper Notebooks vs. Mobile Devices: Brain Activation Differences During Memory Retrieval," *Frontiers in Behavioral Neuroscience* 15 (2021): 34, https://doi.org/10.3389/fnbeh.2021.634158.
39. Hadassah Littman-Ovadia, "Doing–Being and Relationship–Solitude: A Proposed Model for a Balanced Life," *Journal of Happiness Studies* 20 (2019): 1953–71, https://doi.org/10.1007/s10902-018-0018-8.
40. Cassandra L. Brown et al., "Social Activity and Cognitive Functioning Over Time: A Coordinated Analysis of Four Longitudinal Studies," *Journal of Aging Research* (2012), https://doi.org/10.1155/2012/287438; Michael D. Rugg and Alexa M. Morcom, "The Relationship Between Brain Activity, Cognitive Performance, and Aging: The Case of Memory," *Cognitive Neuroscience of Aging: Linking Cognitive and Cerebral Aging* (2005): 132–54, https://doi.org/10.1093/acprof:oso/9780195156744.003.0006.
41. "Harnessing Neurogenesis: How To Change the Adult Brain Through Meditation," EOC Institute, accessed July 20, 2023, https://eocinstitute.org/meditation/the-neurogenesis-guide-how-meditation-changes-the-adult-brain/.

The EOC Institute provides considerable information on the brain, including referenced information on neurogenesis.
42. Abraham M. Joshua, "Neuroplasticity," in *Physiotherapy for Adult Neurological Conditions*, ed. Abraham M. Joshua (Springer, 2022), 1–30, https://doi.org/10.1007/978-981-19-0209-3_1.
43. Richard J. Davidson and Bruce S. McEwen, "Social Influences on Neuroplasticity: Stress and Interventions to Promote Well-Being," *Nature Neuroscience* 15, no. 5 (2012): 689–95, https://doi.org/10.1038%2Fnn.3093.
44. Eddy Larouche, Carol Hudon, and Sonia Goulet, "Potential Benefits of Mindfulness-Based Interventions in Mild Cognitive Impairment and Alzheimer's Disease: An Interdisciplinary Perspective," *Behavioural Brain Research* 276 (2015): 199–212, https://doi.org/10.1016/j.bbr.2014.05.058.
45. Barbara A. Pickut et al., "Mindfulness Based Intervention in Parkinson's Disease Leads to Structural Brain Changes on MRI: A Randomized Controlled Longitudinal Trial," *Clinical Neurology and Neurosurgery* 115, no. 12 (2013): 2419–25, https://doi.org/10.1016/j.clineuro.2013.10.002.
46. Jonathan T. O. Cavanagh et al., "Psychological Autopsy Studies of Suicide: A Systematic Review," *Psychological Medicine* 33, no. 3 (2003): 395–405, http://dx.doi.org/10.1017/S0033291702006943.
47. Raffaella Calati et al., "Suicidal Thoughts and Behaviors and Social Isolation: A Narrative Review of the Literature," *Journal of Affective Disorders* 245 (2019): 653–67, https://doi.org/10.1016/j.jad.2018.11.022.
48. Joanna Kruk et al., "Psychological Stress and Cellular Aging in Cancer: A Meta-Analysis," *Oxidative Medicine and Cellular Longevity* (2019): 1270397, https://doi.org/10.1155/2019/1270397; Shirui Dai et al., "Chronic Stress Promotes Cancer Development," *Frontiers in Oncology* 10 (2020): 1492, https://doi.org/10.3389/fonc.2020.01492.
49. Myrthala Moreno-Smith, Susan K. Lutgendorf, and Anil K. Sood, "Impact of Stress on Cancer Metastasis," *Future Oncology* 6, no. 12 (2010): 1863–81, https://doi.org/10.2217%2Ffon.10.142.
50. Joel E. Dimsdale, "Psychological Stress and Cardiovascular Disease," *Journal of the American College of Cardiology* 51, no. 13 (2008): 1237–46, https://doi.org/10.1016%2Fj.jacc.2007.12.024; Tawseef Dar et al., "Psychosocial Stress and Cardiovascular Disease," *Current Treatment Options in Cardiovascular Medicine* 21 (2019): 1–17, https://doi.org/10.1007/s11936-019-0724-5.
51. Annika Rosengren et al., "Association of Psychosocial Risk Factors with Risk of Acute Myocardial Infarction in 11,119 Cases and 13,648 Controls from 52 Countries (The INTERHEART Study): Case-Control Study," *The Lancet* 364, no. 9438 (2004): 953–62, https://doi.org/10.1016/s0140-6736(04)17019-0.
52. Irving Kirsch and Guy Sapirstein, "Listening to Prozac but Hearing Placebo: A Meta-Analysis of Antidepressant Medications," *Prevention & Treatment* 1, no. 2 (1999), https://doi.org/10.1037/1522-3736.1.1.12a.
53. "Placebo Effect and the Treatment of Parkinson's Disease," Michigan Parkinson Foundation, accessed July 20, 2023, https://parkinsonsmi.org/treatment/entry/placebo-

effect-and-the-treatment-of-parkinsons-disease. Please refer to this page on the Michigan Parkinson Foundation website for more information on the placebo effect on Parkinson's disease.
54. "Mayo Clinic Study Finds Optimists Report a Higher Quality of Life Than Pessimists," *Mayo Clinic*, August 13, 2002, https://www.sciencedaily.com/releases/2002/08/020813071621.htm; Toshihiko Maruta et al., "Optimists vs Pessimists: Survival Rate Among Medical Patients over a 30-Year Period," in *Mayo Clinic Proceedings* 75, no. 2 (2000): 140–43, https://doi.org/10.4065/75.2.140; Becca R. Levy et al., "Longevity Increased by Positive Self-Perceptions of Aging," *Journal of Personality and Social Psychology* 83, no. 2 (2002): 261–70, https://doi.org/10.1037/0022-3514.83.2.261. These references provide more information on the effects of positive thinking on health.
55. Ahmed Tawakol et al., "Relation Between Resting Amygdalar Activity and Cardiovascular Events: A Longitudinal and Cohort Study," *The Lancet* 389, no. 10071 (2017): 834–45, https://doi.org/10.1016/S0140-6736(16)31714-7; Dai et al., "Chronic Stress." These references provide more information on the connection between stress and physical diseases.

5
LEADERSHIP

Introduction

If you are reading this chapter, you probably have an interest in leadership. Your desire might be to become a better leader yourself, or maybe you want to be better at assessing others who claim to be leaders, like your boss. The uncomfortable reality is that leadership is difficult and rare. It is likely that your boss is only a manager, not a leader. This chapter will help you decide.

As firefighters, we know the difference between a fire engine and ladder truck, but the average person probably does not. Fire engines can have certain characteristics that make them different from each other such as the manufacturer or color, both of which are choices of style. However, a fire engine is required to have a necessary and sufficient set of components as outlined by the National Fire Protection Association (NFPA) in *NFPA 1901: Standard for Automotive Fire Apparatus* (fig. 5–1).

The components can look different and come from different manufacturers, but it is not a fire engine unless it has those components. Notice that a ladder

Fire engine (NFPA 1901)	Ladder truck (NFPA 1901)
	Aerial device
Equipment storage—40 ft^3 or greater	Equipment storage—40 ft^3 or greater
Equipment supplied by contractor	Equipment supplied by contractor
Minor equipment	Minor equipment
Water tank—300 gallons or greater	Optional
Fire pump—750 gpm or greater	Optional
Hose storage	Optional

FIGURE 5–1. Necessary and sufficient components of a fire engine and ladder truck in NFPA 1901

truck, according to the NFPA, can have all of the same components as a fire engine, but it does not require the pump, water tank, and hose. However, a fire engine is not a fire engine unless it has all of those components listed. Additionally, there is one item that immediately distinguishes the two different types of apparatus—the aerial device.

Just like the standards for fire engines and ladder trucks listed in the NFPA 1901, leadership has necessary and sufficient components. Just like how fire engines can come in different styles, there may also be different styles to leadership. And, just like how you can immediately differentiate a ladder truck from a fire engine, there is a way to immediately differentiate management from leadership.

The first part of this chapter begins with a defense for the necessary and sufficient leadership components. They are objective and assessable components. *Part 1: Nature and Nurture* explains what cannot be changed for leadership to occur, as well as what can change. What can be changed for leadership is where style and differences come into play. Leadership can be done in different ways, but the necessary components must still be there. *Part 2: Deficiency and Excess* will consider how leadership can be deficient or excessive at supporting the evolution of others. This section will expand on the different ways that the leadership components can be approached. Lastly, *Part 3: Micro and Macro* will examine how leadership transitions from the micro (individual) to the macro (organization) and will present considerations for aspiring organizational leaders.

Part 1: Nature and Nurture

What You Cannot Change (Nature)

The topic of leadership is often framed as an art and not a science. This may be because leadership has historically meant different things to different people. However, in the same way that we can differentiate a fire engine from a ladder truck, it should be possible to differentiate leadership from management. Additionally, even though one fire engine can be different from another fire engine in many ways, there needs to be a minimum set of necessary and sufficient components to make it a fire engine.

The word leadership can be used as a verb and noun. As a verb, *leadership* refers to action. As a noun, *leadership* refers to those in charge. The problem is that *managing* is also an action, and *management* also refers to those in charge. Similar to differentiating between a fire engine and a ladder truck, in order to define leadership, we have to differentiate it from management. As nouns, there is no need to explore the differences since both words refer to people in charge. If no difference exists as verbs, then the two words are merely synonyms. But, the two words differ as verbs, and that means leadership training is not the same as management training. To build the foundation for defining leadership as a verb, we need to first understand what it means to lead. Table 5–1 displays four definitions of the word lead, as well as four definitions of manage.

TABLE 5–1. Definitions of the words lead and manage from Merriam-Webster

Lead – Definition[1]	Manage – Definition[2]
1. To guide on a way especially by going in advance (e.g., led the officers to his hiding place)	1. To handle or direct with a degree of skill
2. To go through (e.g., lead a quiet life)	2. To direct the professional career of
3. To direct the operations, activity, or performance of (e.g., lead an orchestra)	3. To succeed in accomplishing
4. To bring to some conclusion or condition	4. To work upon or try to alter for a purpose

Merriam-Webster categorizes both lead and manage as transitive verbs. If a verb describes action, a transitive verb describes action on an object. So, when lead or manage are used as transitive verbs, they relate to someone acting on another person or thing.

While clear similarities exist between the two words, a significant difference is seen in lead's first definition, "to guide on a way especially by going in advance." The definition's example of someone who "led the officers to his hiding place" indicates someone with knowledge (knowing a hiding place) passing on that knowledge to other people (the officers).

One major point of confusion for some people is about how to classify a person's actions if their title is lead or leader. Those titles are often given to people in the workforce when they are in charge. However, someone can be in charge and not be leading. A manager is also in charge, so is that person managing or leading? If you look into Merriam-Webster's definition of manage, the first example of manage as a transitive verb is to "exercise executive, administrative, and supervisory direction." The first three synonyms for manage in Merriam-Webster are conduct, control, and direct. We often call people "leaders," when in fact they are simply managing.

Like fire engines and ladder trucks, there is some overlap between leadership and management. The Harvard Business Review article "Three Differences Between Managers and Leaders" offers a summary of the differences between the two ideas. The author states that managing involves controlling a group or entities to accomplish a goal, whereas leadership involves a person's ability to influence, motivate, or enable others to contribute toward organizational success.[3] A discussion on why influence is a problematic word occurs later in this chapter. However, two key differences in the article stand out:

- Management is about control, whereas leadership is about empowerment.
- Management can involve either people or entities (e.g., nonhuman resources and processes), whereas leadership involves only people.

Figure 5–2 shows the proposed necessary and sufficient components of leadership. If any one of these components are missing, the action cannot be leadership. A comparison to management is provided within this section, and an algorithm for leadership is provided at the end of this section. The following is the defense for each of the necessary and sufficient components for leadership.

Leadership
Involves humans
Change toward a goal
Change is positive
Change reduces knowledge gap
Change is voluntary
Change lasts in the leader's absence

FIGURE 5–2. Necessary and sufficient components of leadership

Leadership involves humans

> Example: Firefighter Fuego has just been promoted to Lieutenant. Lt. Fuego was hoping to be a company officer but instead was assigned to be in charge of equipment. He can still practice leadership, but Lt. Fuego has to find humans to lead now instead of having humans assigned to him.

Because leadership is about an action on an individual or group, a clear understanding of humans is necessary. In Merriam-Webster, the definition of a human is a bipedal primate mammal.[4] But leadership is not just one mammal acting on another. Interestingly, the Merriam-Webster definition of health is the condition of being sound in body, mind, and spirit.[5] The body-mind-soul triad is a common way to describe the different dimensions unique to humans compared to other mammals. This triad can help you understand leadership better because it suggests that one person can engage with another individual or group in three different ways. If the idea of the body-mind-soul triad is unacceptable to you, feel free to use the bio-psycho-social triad instead.[6] By removing the spirit or soul, however, we reduce human qualities like love and forgiveness to actions that relate only to the mind.

You may be willing to consider a spiritual dimension, but are unsure as to what that means and whether the spirit or soul is just a human trait or something that is eternal, as some religions profess. That argument is beyond the scope of this chapter, but it is worth noting that love and forgiveness seem to be uniquely human actions. Humans love and forgive humans that are from other families and cultures, and even love and forgive different species (e.g., dogs, cats) as well as inanimate objects or ideas (e.g., paintings, books, theories).

On the other hand, maybe the spirit or soul is just an aspect of the mind that is emotional. Mammals have a limbic brain which gives them the ability to care for their kin. However, we rarely see wild or domesticated animals loving and forgiving other animals outside of their kin or different species. Maybe the spirit or soul is just an aspect of the mind that is rational. However, we know that rationality alone does not cause us to love and forgive others. For example, it is rational to not forgive someone who murdered a loved one, or even a former lover who committed an act of betrayal, but people do it anyway. Perhaps the spirit consists of both the emotional and rational parts of the human mind, which nonetheless allows humans to love and forgive in a unique way.

Regardless of your personal stance, for the purposes of this book, a person will be separated into the dimensions of the body, mind, and soul. Whether a soul exists or not, there is no argument that the three original professions were medicine, law, and theology. Perhaps those three professions were the original societal leaders of the body, mind, and soul. Therefore, a leader must effect some action in the body, mind, or soul of an individual or group.

Leadership requires change toward a goal

> Example: Lt. Fuego discovered that newly purchased fire engines needed more frequent repairs because drivers were not operating or maintaining them properly. He realized the engine drivers' behavior needed to change in order for maintenance costs to return to normal.

Leadership and management are actions that both work to accomplish a goal. However, managing a person, thing, or process does not always require change. Managers often simply ensure that a goal is accomplished the same way it was accomplished before. Leaders are identified specifically because of change that they have brought about in the past. Most of the time managers are keeping things stable, whereas leaders are bringing about change.

To be fair, change can occur from management, and stability can occur from leadership. Whether the change occurs from management or leadership, when the change involves people, some aspect of the body, mind, or soul must change. Change can occur in different ways within these dimensions. Someone who is in charge of other people might be called a leader or manager. However, if they are not bringing about change, then they are simply overseeing the work that others are doing. In that case, the word leader is just being used to describe a position or title (noun-based form of leader).

Leadership as a verb involves both humans and change. However, leadership is not about change for the sake of change alone. Change without a goal can happen through exploration or experience, but those instances do not require a leader. It is also problematic to suggest that someone should follow another person without a goal that appeals to the follower. The leader and the individual or group must have goals that align. Therefore, a leader must effect some change in the body, mind, or soul of an individual or group, toward an established goal.

The change must be positive

> Example: Lt. Fuego understood that if drivers continued doing the same thing, nothing would improve. He realized that the drivers needed to add new skills to what they already did, and that the new skills needed to be directly related to the operational and maintenance demands of the engines.

Change can be positive (e.g., improved fitness), negative (e.g., new rules that are unjust), or neutral (e.g., uniform style). Management can implement some of these changes; in fact, a type of management exists called "change management." The difference between change from management and change from leadership is that leadership causes purposeful positive change. It is important to note that assessing change can be objective or subjective. An objective assessment of change considers a new status compared to a baseline status. A subjective assessment of change offers a judgement on the value of change. Change can be objectively positive, but subjectively negative.

Negative change
Leadership does not result in negative changes. One example of negative change is when someone causes another person's body, mind, or soul to change for the worse. For instance, a shift officer might think they are providing good advice by getting their entire shift to adhere to a strict vegan diet or a meat-only diet. However, if one or more of the shift members becomes too sick because of the diet, the result not only negatively impacts the sick individual's body, but also weakens the entire shift. Additionally, if the individual or shift adopted false information, then the mind of the individual or shift was negatively changed as well.

Negative change can also impact the broader community. For instance, if a shift officer succeeded in getting his entire shift to be physically stronger (objectively positive), but the shift used that new strength to start fights, then that would be destructive to the community (subjectively negative). Similarly, a paramedic educating his shift mates on how to start IVs to help work off hangovers while on duty would destroy the public's trust of the fire department.

Accidental positive change
Both management and leadership actions can cause positive change. However, positive change can and does occur without the involvement of a leader or a manager. For example, as a child grows into adulthood, they will be able to jump higher and run faster. Parental involvement is not necessary for those positive changes to occur. It is common to see someone in charge take credit for good things that happened on their watch. But, if there was no goal and no corresponding plan, the positive things cannot be tied to that person.

Alternatively, a positive change can happen to a person, group, or society even if it was not the intended goal. For example, sprinting and long-distance running are two separate activities and require two different types of training and development. If the goal was to make someone better at long-distance running but their sprinting improved in addition to long distance running, then a positive change accidentally took place that was not a stated goal.

Purposeful positive change

In order to know if positive change was made, a goal must be established that is better than the current state or baseline. The goal demonstrates that the action has a purpose, and the pursuit of that goal demonstrates that an individual or group believes that the goal is worthy. Without an established goal, any positive change to a person, group, or society must be considered an accident, and leadership does not occur by accident. Additionally, without a plan to achieve the goal, there is no way to give credit to someone who is claiming to have caused the change. However, even with a goal and a plan, it is necessary to account for positive change that may have occurred anyway, like a child's physical growth.

It can take time to determine whether objective positive changes to an individual or group were subjectively positive. Ideas, societal norms, and science can change over time, and what may be considered subjectively positive in one era is not always considered positive the next. However, the changing of ideas or science does not mean objective positive change is impossible. Long-distance running and sprinting are two physical skills that require different training. We can disagree on which one is better, but becoming better at either one is objectively positive and can be documented.

Leaders do not make neutral or negative changes. Neutral and negative changes to an individual or group provide no benefit from baseline. Furthermore, any negative change to an individual or group would require involuntary action or the use of incentives, because no sane individual or group would voluntarily negatively impact an aspect of their body, mind, or soul without believing the change would result in a net benefit to one or more of their dimensions. Organizations do not train people to become leaders in hopes that they may make neutral or negative changes. Leaders make positive changes. Leaders must improve individuals and at least not harm society. Therefore, a leader must effect positive change in the body, mind, or soul of an individual or group toward an established goal.

The change must reduce a gap in knowledge

> Example: Lt. Fuego received permission to create a training program for the drivers that were assigned to the new engines. He knew they were all experienced drivers, so the training only needed to focus on the new skills required for the new engines and not on the basic engine driving knowledge.

In the fire service, you hear people say "bad officer," not "bad leader." Bad leader is an oxymoron. Someone who knowingly provides false information to another is attempting to deceive that person, and that is not leadership. Someone who mistakenly or ignorantly provides false information to another is not leading because they don't actually understand what is happening. In both cases, an actual knowledge gap was not closed because the information was false.

Without knowledge to pass on to someone else, the most you can say about a person who was trying to lead another person or group is that they were co-exploring. Without the passing on of knowledge, the leader cannot expect any change to occur in the follower, and therefore the leader-follower relationship is not necessary. However, even when someone has knowledge, change can occur that has nothing to do with leadership, even if the change is positive. Positive change without leadership can occur when you do something to or for someone else, when you force them to change, or when you use different types of influence to get someone to change.

It is not economic exchange
You engage in economic activity all the time. The basis for economic exchange is that you believe you are ending up with a positive outcome in exchange for something of value. Having positive change done to you or for you, whether you are paying for it or it's free, is a type of economic exchange, not leadership. As examples;

- **Body**: A dentist gives a patient a filling
- **Mind**: A lawyer updates your written will after a new child
- **Soul**: A spiritual leader declares your soul has been cleansed

It is not forced action
Changing behavior can be done by force through controlling resources or the environment. For example, if you lock an unhealthy eater in a room for days and only give them healthy food, they will eventually eat healthy food. If you put a person in a maze, you are literally restricting the directions in which they can walk and thus the potential outcome. Positive change by controlling the environment or resources is management science, not leadership.

It is not using influence
Changing behavior can also be done through influence. A person can influence someone to do something because of their power, sexuality, or money, to name a few examples. However, if influence is the key to leadership, then we have to conclude that those with the most power, attractiveness, or money are the

best leaders. That conclusion is problematic for two reasons because it means the following:

1. You cannot be a leader if you are not in a position of power, if you are considered unattractive, or if you are poor.
2. Leadership is not a subject that can be taught, but is a consequence of success that you earn or inherit.

To get people to positively change their behavior without using influence or controlling their environment or resources requires changing the way people think. In order to change the way someone thinks, new knowledge must be taught. For knowledge to transfer from one person to another, one person must know something that the other does not. No change occurs if a person already has the knowledge. Similarly, directing others to do a task that they already know how to do is not leadership. We could call that directing or commanding. In fact, we have algorithms that tell us what to do and when to do it within emergency medical services (EMS) protocol.

Leading requires a gap in knowledge between the leader and follower, which is then reduced through the interaction between the two. Therefore, a leader must effect positive change in the body, mind, or soul of an individual or group toward an established goal by reducing a knowledge gap.

The change must be voluntary

> Example: Lt. Fuego's training program became mandatory for drivers of the new engines. Lt. Fuego developed and delivered the class and word spread that Lt. Fuego was a very good engine driver. Firefighters began voluntarily asking him for training on preparing to be a driver.

Up until now, the components of leadership could also be applied to management. While management can involve things or processes, it can also involve humans. While management can involve stability or even neutral or negative change, it can also cause positive change toward established goals. Management can even be used to reduce a knowledge gap in others. Figure 5–3 compares leadership and management components.

Consider that in a fire academy, the instructors are causing positive change in humans by reducing knowledge gaps. For example, the instructors may have goals for improving trainee physical fitness, basic firefighting skills, and basic

Leadership	Management
Involves humans	May involve humans
Change toward a goal	Change may be goal oriented
Change is positive	Change may be positive
Change reduces a knowledge gap	Change may reduce a knowledge gap
Change is voluntary	Change is involuntary
Change lasts in the leader's absence	Change may last or not

FIGURE 5-3. Comparing leadership and management components

EMS skills. However, if the trainee does not participate in mandatory training, they will be discharged from the academy. There is nothing voluntary about what a trainee must do once they are in the academy. The academy is a controlled environment. Controlling an environment is one way to manage people, but you can also control resources or influence behavior using incentives and disincentives. Using management science negates the idea of followership and treats humans as objects.

Aristotle taught that without voluntary choice, an individual cannot be fully responsible for their actions.[7] Part of voluntary choice involves the ability to do otherwise (choose differently). Management science relies on reducing the freedom of choice by controlling the environment or resources or by using enough incentives or disincentives to control the choices of individuals and groups. It is true that you always have a choice, even if the choice involves your own death. However, when doing leadership, the threshold for freedom of choice ends the moment any control factors are used, no matter how small. A leader firstly allows a person to freely choose whether to receive the knowledge and, secondly, whether to adopt and use the knowledge. Essentially, the leader is helping a person to enter into and complete an evolution of their own self. The use of incentives or disincentives by someone in charge demonstrates their belief that they would not be followed otherwise. Therefore, a leader must effect a voluntary, positive change in the body, mind, or soul of an individual or group, toward an established goal, by reducing a knowledge gap.

It should be noted that leadership is not bound by time or space. For example, if you practice a religion, it means that you follow the guidance of a religious leader who may or may not be alive and who may or may not live near you. While it is true that many people are compelled to stay in the religion they were born into or are even forced into a different religion, those are examples of management because voluntary choice is not being allowed. Similar to religion, a leader of philosophy, politics, business, health, or sports can offer guidance that is followed on a different side of the planet or many years after their death. The key is that the follower has identified guidance from the leader that aligns with a desired positive change and voluntarily chooses to follow that guidance, which reduces a gap in their knowledge.

The change must last in the leader's absence

> Example: After Lt. Fuego's class, the operational and maintenance issues went away and repair costs returned to normal. Rules were put in place to ensure that drivers of the new engines continued the new skills. The firefighters who voluntarily learned continued the new skills without rules.

Time is the final aspect of leadership that must be considered. Is it sufficient that a leader causes positive and voluntary change to the body, mind, or soul, even if that change evaporates when the guidance from the leader ends? It is one thing for a person to voluntarily learn something new from someone else. It is another thing for a person to voluntarily use that knowledge to change their behavior while the person who taught them the new knowledge is still around. It is yet another thing entirely for a person to adopt that behavior change when the leader is absent. The longer the change lasts in the absence of the leader, the more we can conclude that the follower voluntarily adopted and valued the change.

Consider the change that occurred because of Lt. Fuego's class. Experienced engine operators learned new skills that were required for the new fire engines. Some behavior change needed to occur to the engine operators' body and mind, which was positively related to some goal. However, the course was mandatory, which made it management. Additionally, rules were put in place to ensure that the engine operators continued using the new skills they were taught. The behavior change lasted, but only because of management.

Now consider the firefighters that volunteered to learn from Lt. Fuego after hearing about his knowledge and skills as an engine operator. The firefighters who learned from Lt. Fuego voluntarily wanted to learn what the engine operators were forced to learn. Lt. Fuego did not have to force or incentivize them to learn these new skills. Additionally, they continued to practice their skills without mandates. This type of lasting behavior change caused by Lt. Fuego is leadership.

As a thought experiment, it is possible that one or more of the firefighters who voluntarily learned from Lt. Fuego stopped practicing those skills at some point. That would indicate that they rejected the guidance that caused the change and no longer want to follow Lt. Fuego. The amount of time that a change lasts in the absence of the leader indicates whether an individual or organization has adopted or rejected the change. Time also acts as a good indicator for measuring the impact of a leader. Figure 5–4 provides an example of calculating a relatively big change that lasted for six months and then stopped, compared to a relatively small change that continued over the course of three years.

The equation in figure 5–4 is a theoretical formula for quantifying leadership. In chapter 1, we explained that resilience is a measurement of how much challenge the body, mind, and soul can overcome. Leadership is all about increasing the ability of someone to overcome a new challenge through a voluntary exchange of knowledge. The formula states that leadership equals the change in the body times the change in the mind times the change in the soul times the duration of time these changes lasted for. This formula could be applied to change caused by management or leadership, so the user has to assess whether the changes involved an established goal and were voluntary. This equation could also be used to assess either positive or negative change. But, in order for the change to be attributed to leadership, the change would need to be positive.

> Figure 5–4 depicts two hypothetical situations. Person 1 made positive change in their body by 20%, their mind by 50%, and their soul by 30% over the course of 6 months. Person 2 made positive change in their body by 10%, their mind by 10%, and their soul by 10% over the course of 36 months. Person 1 had a total change of 14.04 and Person 2 had a total change of 47.91.

$$L = (1 + \Delta B) \times (1 + \Delta M) \times (1 + \Delta S) \times t$$

Leadership = Change in body above baseline ×
Change in mind above baseline ×
Change in soul above baseline ×
Time

Example
Person 1: $14.04 = (1 + 0.2) \times (1 + 0.5) \times (1 + 0.3) \times 6$
Person 2: $47.91 = (1 + 0.1) \times (1 + 0.1) \times (1 + 0.1) \times 36$

FIGURE 5–4. A theoretical equation for quantifying leadership

The amount of time a voluntary change lasts indicates how valuable the leadership guidance was to the person who received it. When a person freely chooses to act in a way that discontinues or diminishes the positive changes that had been adopted, the person is rejecting the guidance of the leader. This might occur because the person no longer values the guidance, or because they have adopted different guidance that they value more. If no active follower has adopted the guidance of the leader and continued the behavior on their own, then leadership has not occurred. Therefore, a leader must effect a voluntary, lasting, positive change in the body, mind, or soul of an individual or group toward an established goal by reducing a knowledge gap.

Summary

We have now outlined both the necessary and sufficient components of leadership:

1. Leadership requires engaging with humans, not things or processes. Human dimensions conventionally include a body, mind, and soul.
2. Leadership requires changing some aspect of at least one human dimension of an individual or group.
3. Leaders are expected to bring about positive change. Different types of positive change exist, and sometimes they can be at odds with each other, like with sprinting and long-distance running. Regardless, positive changes to the body, mind, or soul are measurable. Sometimes, it can take years to determine whether a change was positive or negative to society, which means that determining leadership on a larger scale is not always possible immediately.
4. Leaders must have knowledge that the follower does not have and pass their knowledge on.
5. Leading requires voluntary participation from the follower. This is a major distinction from management, which uses authority, incentives, or disincentives to change behavior.
6. Any change must be shown to last in the leader's absence, which demonstrates that the knowledge has not only been successfully passed on to the follower, but has been freely adopted.

Leadership and management can produce the same change in behavior (fig. 5–5). However, the two methods use different ways to achieve the same outcome. Management involves controlling people's choices—even if the change is positive—toward an established goal and reduces a knowledge gap. Leadership, however, makes a voluntary, lasting, positive change to the body, mind, or soul of an individual or group toward an established goal by reducing a knowledge gap.

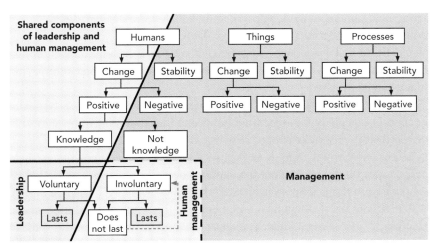

FIGURE 5-5. An algorithm for getting to leadership

***

The previous definition of leadership that is put into sentence-form is admittedly long and cumbersome. That is because this definition simply strings together the necessary and sufficient components of leadership. However, leadership is such an important topic that it deserves to have a definition that is easier to remember and say to someone. For example, even though a fire engine has a list of necessary and sufficient components outlined by the NFPA, you don't tell people that a fire engine is the entire list of components. The purpose of a fire engine is to put out fires. So, you might say that a fire engine is like a toolbox on wheels that can pump water to a fire. Obviously that sentence does not capture all of what a fire engine is or is not, but it is probably an acceptable definition for most people, down to the age of an elementary school child.

Leadership can also have a shortened definition if it properly captures the essence of what leadership is. When something has a list of necessary and sufficient components, each component is equally important. However, just as it may not be necessary to tell a child that a fire engine must have a siren, it is also probably not necessary to explain to a child that leadership must last in the absence of the leader. The purpose of leadership is to help humans be more capable of overcoming challenges—to be more resilient. At its core, then, *leadership is supporting the evolution of another person's body, mind, or soul.*

What You Can Change (Nurture)

Changing human behavior is not easy, especially if the change involves freedom from control techniques. Truly, any change that a leader helps to usher in for another human actually helps that human to evolve a better way of being, thinking, or interacting (body, mind, or soul). By using the objective and precise definition of leadership from the first section, it becomes apparent that leadership is hard to accomplish and is probably more rare than common, but that it can be assessed.

It might sound wonderful in theory to wish that no one would ever be forced to change, especially for something important like your job. However, people are not always willing to adopt a behavior change voluntarily, so management is necessary at times. This is especially true for organizations that are held accountable by stakeholders or the public.

There are three areas of leadership that you can change. You can change which dimension of the human you focus on—body, mind, or soul. You can also change the way in which you close the knowledge gap. Lastly, you can change the way you motivate someone to voluntarily engage in the evolution of their behavior.

Human dimensions

The first area of leadership that you can change is the human dimension that you focus on. In the fire service, many different disciplines exist for you to choose from. Examples include engine work, truck work, EMS, fire investigation and code enforcement, special operations, and health and wellness. Each firefighter has a different interest, and throughout your career you will likely gravitate to a specific area of focus. Perhaps, eventually, you will get good enough to teach others about that area if you are not already doing so.

Leadership is similar. Because there are three human dimensions, you will likely gravitate toward a dimension that most interests you. The fact that you are a firefighter, an emergency medical technician (EMT), or both indicates that you already have some interest in the human body dimension. That could include your interest in a physically challenging job or your interest in providing life-saving care to members of your community. If you have taken an assignment in education and training or peer support for your department, you probably have an interest in the human mind dimension. Perhaps you have considered chaplaincy or you facilitate a religious study group. That would indicate your interest in the human soul dimension.

Closing the knowledge gap

The second area of leadership that you can change is the method by which you reduce a gap in knowledge for the person you are helping to evolve. There are three different methods for transferring knowledge to another person. One method is experimentation. A second method is observation. A third method is education.

As a leader, you might prefer one method of knowledge transfer over another. Each method has its strengths and weaknesses. It is your job as a leader to determine which method you are best at and which method is best suited for supporting the evolution of the person you are leading.

Motivation

The third area of leadership that you can change is the way in which you motivate others to evolve. Greek philosophers identified two ways in which humans are motivated. One is through hedonia (pleasure seeking).[8] The other is through eudaemonia (fulfillment of potential).[9] In order to be engaging in leadership, you have to focus on eudaemonia as a motivational tool, but there are different ways to do that. This book covers three ways to motivate others that were made famous by Daniel Pink. The first is by supporting the individual's autonomy. The second is by offering a path to mastery. The third is by highlighting the connection to a sense of purpose.

Part 2: Deficiency and Excess

Human Dimensions

Humans are unique not only because of genetics, but also because of the different paths that individuals take in how they evolve their body, mind, and soul. Free will is one of the main differentiating aspects between animals and humans. Besides genes, it is the freedom to choose that makes you different from others and makes the world so diverse.

The definition of leadership presented in this book does not require you to help someone evolve in all three human dimensions (fig. 5–6). However, humans are not limited to evolution in one dimension, and therefore leadership is not limited to one dimension. Just as it is possible for a firefighter to master multiple areas of expertise, leaders are certainly capable of doing leadership in

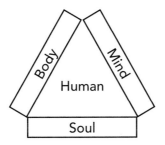

FIGURE 5–6. Change can occur in one of three human dimensions.

more than one human dimension. That means that, with regard to the human dimensions, a deficiency in leadership means failing to help someone evolve in at least one dimension. If no change occurs in the human you are trying to help, then that interaction was not leadership. On the other hand, it is excessive to try to focus on all three human dimensions all the time with everyone you are trying to help.

It should be noted that it is absolutely not necessary to help others to change in every, or any, interaction. Interactions can, and should, happen all the time between humans that do not involve changing one another or attempting to help the other person to evolve. Those interactions are a necessary aspect of human interaction and help to build families and communities. The point is, if you are trying to do leadership but you don't help the person evolve in some way, then that interaction cannot be considered leadership; otherwise, every interaction would qualify as leadership.

Leading others in their evolution to become better humans takes time and energy. Recall from chapter 1 that resiliency describes a person's ability to overcome challenges to the body, mind, or soul, and that growth in any one of those areas does not happen immediately. Ultimately, leadership aims to make humans more resilient so that they can overcome more challenges after interacting with you. If leadership requires helping someone to evolve in at least one dimension, then perhaps a good leader is capable of helping others to evolve in at least two dimensions. Perhaps the greatest leaders are capable of helping other humans to evolve in all three human dimensions.

Because leadership deals with humans and humans have three dimensions, then by definition, there will be diversity involved in leadership. Toolboxes have different tools to address different issues. Diversity in a toolbox is about the different tools that you have, not about having an entire box of the same tool in different colors of paint. However, it is the case that you may need several types of a specific tool that are all slightly different.

Fire departments have different types of apparatus (e.g., fire engines, aerials, rescue squads, and ambulances) to address different needs. But not every fire department needs the same type of apparatus. Fire departments for cities, wildland, and airports will all need fire engines with different specifications. Even within one fire department, there may be different types of fire engines to address different needs. However, they are all still fire engines if they meet the necessary requirements specified in NFPA 1901, just different types.

Having an entire fleet of fire engines with the exact same specifications, just painted in different colors may be aesthetically enjoyable, but it doesn't help the department to address different needs. It is certainly the case that aesthetics and function can coexist. You can have a tool box full of different tools that also happen to be painted different colors. For instance, you can have an orchard of apple, pear, and peach trees, and you can even have different types of apples, different types of pears, and different types of peaches. But if you replace the fruit trees in an orchard with cherry blossom trees just because the trees are beautiful to look at, then that means you have reduced the function of the orchard. If you sacrifice function for aesthetics, then eventually the desired function can no longer be carried out.

Leadership will also come in different packages for different human dimensions, but it is still leadership if it meets the necessary and sufficient components. Just as humans have three different dimensions and you can provide leadership for any one, each dimension has different areas of focus (fig. 5–7). As examples, you can focus on activity and nutrition for the body dimension, learning and motivation for the mind dimension, and love and forgiveness for the soul dimension. Furthermore, each area of focus can then be broken down into sub-areas. As areas of focus become more granular, the leadership requires different, more specialized knowledge. Diversity in leadership is necessary to address different human dimensions. However, the diversity has to involve the functionality of leadership, whether it is aesthetically pleasing or not.

Leadership and diversity are inherently connected. A city fire engine is not the right tool for wildland fire, even if both fire engines meet NFPA 1901. For

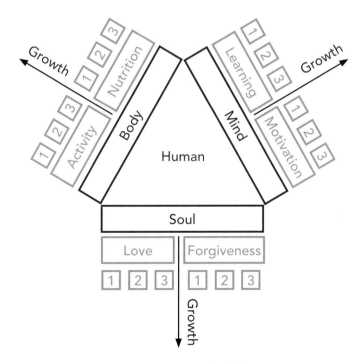

FIGURE 5–7. Each human dimension can change in different ways.

a wildland fire, the fire engine probably needs a different suspension and different tires, and perhaps it has four-wheel drive. There are different tools for different jobs, different fire engines for different environments, and different leaders for different human dimensions.

Even within the same dimension, a different leader may be needed. Fire engines come in different shapes and sizes and with different wheelbases and tank sizes. Those differences do not make it less of a fire engine, just better suited for a certain need. Similarly, different leaders are often needed even when dealing with the same human dimension. Leadership is not one-size-fits-all, and there is room for many types of leaders.

The body dimension

The first human dimension is the body. The human body is the most foundational dimension. The body allows the brain to survive and is where the soul resides. But the body is not simply a piece of stone that is carved and polished like a statue. It is a system of systems and is very complex. Recall from chapter 2 that three different physical activities can affect your body (exercise, fasting, and sleep), and each one is a topic that has been extensively written about by expert specialists. Additionally, chapter 3 details the different aspects of nutrition that you can change (nutrient variety, nutrition density, and nutrient quantity), and each nutritional aspect is its own specialization.

A leader who can help someone evolve with their fitness may not be the right leader to help that person evolve in their nutritional habits. Even within fitness or nutrition, one leader may be unfamiliar with certain specialty areas, and so a different leader would be better suited to address that specific issue. Similarly, even though the brain is part of the human body and is made healthier with exercise, it is often treated separately from just an organ with blood flow because the brain is where the mind dimension rests.

The mind dimension

The second human dimension is the mind. The mind is the seat of our consciousness and is the part of you that can knowingly connect with another person or with the environment. The mind is the part of you that thinks about things and engages in new learning and motivation. Recall from chapter 4 that your physical brain has an impact on the way you think. However, the mind also plays a critical role with how you behave and thus how your brain and even your genes function.

A leader who can help someone learn new things may not be the right leader to help that same person be motivated. Even within learning and motivation, a leader may specialize in different areas of focus. Humans learn in different ways, and as a leader, you can choose to help others learn through experimentation, observation, or education. If you are motivating others, you can choose to focus on autonomy, mastery, or purpose. But even the best leader who helps to increase motivation in a philosophical or a secular way may not be able to reach someone who is seeking spiritual guidance.

The soul dimension
The third human dimension is the soul. The soul is not something that is classifiable with science—at least, not yet. However, for thousands of years humans have talked, sung, written, and fought about the connection of something inside of humans to something eternal and everlasting. The fire service has always faced matters of the soul because it has always faced matters of life and death. That does not mean that everyone in the fire service recognizes a soul. However, the fire service has a proud history of honoring fallen firefighters and their families, and that history is heavily intertwined with religion and spirituality.

The soul is similar to the body and mind dimensions in that it has different areas of focus. Love and forgiveness are two different areas that may require two different leaders. It is possible to grow your ability to love others without ever having had the need to forgive on a major issue. As a leader, you may have the knowledge and experience on forgiving a specific issue that is exactly what someone else needs to evolve in that area. Further, there are many different faiths, sects within faiths, and even non-faith-specific spiritual belief systems. Leaders of the soul help others evolve in their ability to love, forgive, and connect with other people and The Eternal. Firefighters who recognize a soul know that the connection with The Eternal is the ultimate connection, and leaders of the soul help to strengthen that connection.

Closing the Knowledge Gap
The second set of options you have as a leader are the methods by which you close the knowledge gap of the person you are helping to evolve. To support the evolution of another person, it is important that you understand two basic strategies. First, you must understand how to assess a person's behavior in order to know the knowledge gap you will work to close (this book uses Aristotle's philosophy on responsibility as a way to assess behavior). Second, you must understand how to support a person's evolution in their behavior by delivering the knowledge they need to be successful in that change.

Aristotle and assessing behavior
Assessing behavior is a matter of breaking decision-making into smaller parts and identifying where there is a need. Certain human behaviors are influenced by instinct. Also, humans are genetically different from one another, and genes

have a strong influence on behavior. However, humans have the capacity for rational thought, and thus free will. It is because of rationality and free will that humans can be considered responsible for their behavior.

Aristotle's teaching on moral responsibility provides a guide for analyzing human decision-making and is still used today. Aristotle taught that, to assign responsibility to someone for their actions, four assessments have to be conducted. The results of the four assessments allow you to determine whether the person is partially or fully responsible for their behavior and whether to assign praise or blame. Figure 5–8 provides a model for assessing change by taking Aristotle's four assessments and making them easy to understand and apply.

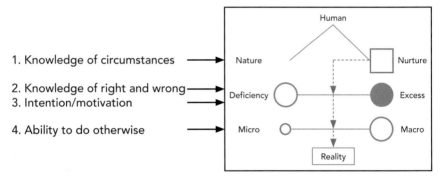

FIGURE 5–8. Change Assessment Model

Knowledge of the circumstances: nature and nurture
Aristotle's first assessment determines whether the individual had knowledge of the circumstances surrounding his behavior. Figure 5–8 reminds the viewer that knowing the circumstances means understanding what cannot be changed (nature) and what can be changed (nurture) about an individual or situation. Even though knowing the circumstances includes understanding what cannot be changed, the graph indicates that the focus should be on what circumstances can be changed.

> Example: Recruit Newbie is at the station for the first time and is cleaning the floors in the bathroom. The station officer has set the radio so that the tones for each station go off so that he can keep up on what is going on throughout the district. The tones for Recruit Newbie's station drop. Does Recruit Newbie understand what that noise was? If so, does he know that there is a way to check whether the tones that drop are for his station or a different station?

Knowledge of right and wrong: deficiency and excess
Aristotle's second assessment determines whether someone had knowledge of right and wrong. Figure 5–8 reminds the viewer that knowing right from wrong means understanding the spectrum of potential choices ranging from deficiency to excess. Aristotle teaches that both deficiency and excess are wrong and that right exists in the virtuous middle.

> Example: Recruit Newbie hears the tones drop and realizes that a call is going out. He does not know for sure if the tones are for his station. Does Recruit Newbie know that if the tone is for his station, no one is going to come looking for him so staying put is deficient? Does he know that sprinting through to halls to the captain's office every time the tones drop is excessive? Does he know a reasonable solution if he is unsure of who the tones are for?

Intention to choose: deficiency and excess
Aristotle's third assessment determines someone's intention regarding their action. Figure 5–8 prompts the viewer to identify where the person's choice landed on the spectrum of deficiency and excess. A wrong act can be done because the individual lacked the mental capacity to make the right choice. For instance, a baby cannot comprehend right or wrong and neither can someone who is unconscious or semiconscious. Additionally, a wrong act can be done by mistake if someone thinks they know the circumstances or know right from wrong, but they actually do not. Unintended actions can even occur if a person is completely conscious, like a muscle twitch. However, someone can have knowledge of right and wrong, be completely conscious, and still act wrongly on purpose. In this situation, a person would need to change their intention, or increase their motivation, to choose right from wrong.

> Example: Recruit Newbie continues to clean the floors in the bathroom after hearing the tones drop. He is fully conscious and has no known mental handicaps that would affect his ability to choose the right action.

Ability to do otherwise: micro and macro
Aristotle's fourth assessment determines whether the person was acting as a free agent or whether some force was affecting the person's ability to choose. The Change Assessment Model reminds the viewer to consider the

relationships between micro and macro, which can help to identify factors that impact decision-making. A sudden injury (micro) can prevent a person from making a right choice even if they want to make the right choice. Family and employers (macro) often use rules, incentives, or disincentives to make someone behave a certain way. Even natural events (macro) can force someone to make a choice they otherwise would not have made. This last assessment does not necessarily excuse someone for making a bad decision if control factors were being used. However, it does mean that the person using the control factors was doing management and not leadership.

Recruit Newbie explained, "The senior man on the shift said to mop all the bathroom floors and not to leave for any reason until they are all done . . . or else!" Recruit Newbie had stayed because he was threatened with some consequence if he left the bathrooms before he was done mopping. Without that disincentive, would Recruit Newbie have responded? What other micro or macro forces could have prevented Recruit Newbie from responding?

> Example: The shift waited for a minute in the fire engine for Recruit Newbie, but the crew became anxious to respond to the call so the officer told the driver to start driving. When the crew got back, the officer called Recruit Newbie into his office. The officer asked Recruit Newbie if he knew which tones dropped and Recruit Newbie responded, "Yes." The officer asked Recruit Newbie if he knew what should be done when the tones for their station drop and again Recruit Newbie responded, "Yes," and explained the correct answer. Finally, the officer asked, "Then explain to me why you did not get on the engine for that last call!"

You can understand a person's reality after considering these four areas. A person's reality represents their knowledge of the things they cannot change and things they can, their knowledge of how to navigate the situations that they are faced with, the actions they chose, and the forces that influenced their actions. For example, a firefighter's weight is highly influenced by DNA (nature), but exercise and diet will play a role in their reality. You can use figure 5–8 to assess a firefighter's responsibility and potential areas for change in regard to weight by asking the following questions:

- What does the firefighter know about genetics (nature)? What do they know about the relationship between lifestyle choices and body weight (nurture)?

- What does the firefighter know about how to navigate physical fitness and nutrition choices (deficiency and excess)?
- Is the firefighter making the right choices (deficiency and excess)?
- Is something or someone controlling the firefighter's choices (micro and macro)?

Aristotle's philosophy on responsibility provides a guide to assessing behavior change. You can be held fully responsible for your behavior if you knew the circumstances, knew right from wrong, intentionally acted the way you did, and were not controlled by some external force. In order to change your reality, you need to change one or more of the components of responsibility. Therefore, a leader must assess which of these factors is causing the specific behavior and help the individual to see that.

Assessing the behavior change gap

Assessing behavior change is a matter of identifying the gap between two realities: desired and current. The desired reality is an individual's goal, and the current reality is the individual's baseline. Assessing behavior change includes identifying a goal, establishing a baseline, and identifying the required learning needed to close the gap.

Identify a goal

The first step in assessing behavior change is to determine where an individual wants to end up. It may seem out of order to establish a goal for someone before assessing their current situation or baseline. However, consider that leadership is about voluntary change, and not everyone is ready to change or work toward a goal. If someone has no desire to move toward a goal and change their behavior, then there is no need to worry about what their current baseline is. In order to change someone's behavior that has no desire to change, management must be used.

For a person to change their reality voluntarily, they must agree to a goal without coercion. A person who agrees to a goal because their job depends on it or because a reward or punishment is at stake is doing so because of management, not leadership. The SMART goal method provides a systematic way to create goals. SMART goals are Specific, Measurable, Attainable, Relevant, and Time-Bound. SMART goals have a higher chance of being accomplished because they are not deficient or excessive. A goal that is not challenging enough does not stimulate the individual, and a goal that is too challenging will intimidate the individual. The use of the SMART goal model reminds those who are making the goals to keep each goal *attainable* (the "A" in SMART). In order to keep goals attainable, adjustments may have to be made after the baseline has been clearly established.

Establish a baseline
The second step in assessing behavior change is to establish a baseline measurement. The difference between the goal and the baseline is the gap between the desired reality (goal) and the current reality. The reality that is being measured might relate to the individual's body, mind, or soul. The following are examples of aspects of reality that can be measured for each human dimension:

- **Body**: Weight is an aspect of the body which can be measured by using a scale. Another measurement that is related to weight, like body fat percentage, can also be used to quantify change. An individual's body fat may decrease based on exercise and nutrition even as their weight stays the same or even increases due to muscle gain.
- **Mind**: Knowledge of a subject matter is an aspect of the mind that can be measured using a subject matter quiz on facts or concepts. Another way to measure knowledge of subject matter is to demonstrate the ability to carry out a task.
- **Soul**: Forgiveness is an aspect of the soul that can be measured using a questionnaire about a firefighter's feelings toward their shift or supervisor.[10] Another way to measure forgiveness is to observe someone's behaviors toward an individual or group that is accused of having done wrong.

Measuring the body is different than measuring the mind which is different than measuring the soul. However, any targeted change must be measurable in some way, even if it is self-reported by the individual. For example, someone might not feel forgiving one day but may feel very forgiving the next. The measurement offers a snapshot of the current reality being assessed and can change over time. This measurement can even change from day to day.

Once you have obtained a baseline measurement, be sure to reassess the goal. Adjusting the goal may be necessary if it becomes clear that achieving the goal does not follow the SMART guidelines. In other words, you may have to break the overall goal (future reality) into smaller goals in order to align with the *attainable* component of the SMART model.

Identify required learning
The third step in assessing behavior change is to identify new learning that will lead to the new behavior needed to achieve the new reality (goal).

> Learning → Behavior → New Reality (Goal)

You now know that a person's reality is the result of the four factors outlined in Aristotle's assessment of responsibility. Using management science, behavior change can occur simply by creating rules or using incentives or disincentives. Management can also be used to force people to learn new knowledge. However, leadership causes behavior change because of the voluntary adoption of knowledge.

By using Aristotle's four assessments, you can identify potential knowledge gaps. The individual who voluntarily seeks behavior change must fulfill at least one of the following conditions:

- Learn something new about the circumstances
- Learn something new about choosing right from wrong
- Become more motivated to choose right from wrong (which requires learning)
- Learn how to obtain help with overcoming a barrier or controlling factor

Once any knowledge gaps are identified, you can then identify the necessary learning plan and lessons to deliver.

Aristotle taught that an individual might not be fully responsible if their behavior occurred because of some controlling factor. Therefore, it is important to determine whether the person is actually voluntarily choosing to change, or if they are being pressured or forced. This is not always easy to know. For example, consider the following questions:

- If a behavior change is enforced (with rules) for a person to keep their job, and the person then changes their behavior in order to keep their job, is that behavior change voluntary?
- If a person voluntarily changes their behavior (without rules) on the job, but at home behaves differently, has leadership occurred?
- How do you differentiate and account for incentives or disincentives that might be physical, psychological, or spiritual?

It is important to repeat that management is not inherently bad. Management is necessary to enforce certain behaviors because organizations are held accountable by stakeholders. However, it is important to distinguish between management science and leadership. Someone practicing leadership does not put pressure on anyone to follow them. The choice to learn from a leader and adopt their guidance must be voluntary. Any use of incentives or disincentives, no matter how insignificant they may seem, demonstrates that the person who is trying to help someone change their behavior relies on management science. If leaders can use incentives or disincentives, we would have to conclude that the best leaders are the richest or most powerful which, as described earlier, is highly problematic.

Addressing the behavior change gap

Addressing the behavior change gap is a matter of successfully developing a learning plan, delivering the learning plan, and documenting the targeted changes. The learning would be deficient if it does not address at least one of the four assessments from Aristotle. However, it may be excessive to address all four assessments at one time. Additionally, the learning must be tailored specifically to help the individual change their behavior in order to reach their goal.

Recall from the definitions of "lead" from table 5–1 that the first example includes the phrase, "especially by going in advance." The idea of going in advance suggests that the leader has knowledge that a follower does not yet possess. To address the behavior change gap, the leader must devise a plan to reduce the knowledge gap of the follower. The plan can be implemented with the leader and follower physically in the same place, but that is not required. Consider that many great historical thinkers have changed people's lives some tens, hundreds, or even thousands of years later and in many different parts of the world. Now, with phone and internet technology, learning can occur in real time between people who are on opposite sides of the world.

Developing a learning plan

The first step in addressing a behavior change gap is to devise a learning plan. There are entire books written about learning design, and you can even get a college degree in that area. However, a comprehensive review of learning design is not within the scope of this book. Instead, the purpose here is to underscore the three main ways that a person can learn so that, whether you are designing something formal or off-the-cuff at the firehouse, you understand the different options for transferring knowledge to others.

The plan that a leader develops must increase the person's knowledge of the circumstances, knowledge of right and wrong, intention to choose (motivation), or the ability to do otherwise. It is important to note that a person's ability to do otherwise might increase only through assistance, and sometimes they may have no ability to do otherwise. Therefore, if the learning plan addresses a person's ability to do otherwise, it should either include how to access available assistance, how to cope with the inability to do otherwise, or a combination of both.

The leader must ensure that the learning plan for knowledge, motivation, or ability to do otherwise aligns with specific behaviors needed to achieve the voluntary goal. Like the goal, the learning plan must be voluntary. Otherwise, any success in achieving the goal is either due to management or an accident, neither of which is leadership. When developing your learning plan, you can choose between three different ways that people learn:

1. Experimentation
2. Observation
3. Education

Experimentation can be very rewarding, but also very risky. Some of the greatest inventors and scientists use varying degrees of experimentation, either by doing completely random experiments or extremely planned ones. As a leader, you may know that someone needs to experience something before they understand it. While someone can gain a deep understanding through experimentation, it can be a slow and costly way of learning. Additionally, the resources to learn through experimentation are not always available.

An intentional and safe plan can allow a follower to learn something new through experimentation. For instance, practicing rescue techniques with either a weighted training mannequin or a shift mate that is fully geared up can help a firefighter understand the difficulties involved. A firefighter may have to attempt certain methods and even fail at them before adopting a best practice. The phrase "failing forward" captures the idea of learning through experimentation.

Observation occurs all the time and is a very common way to learn. By observing others' behavior, you can learn a potential way of life from others. For specific skills, it may be necessary for a new firefighter to watch a veteran do something in order to do it successfully themselves. Even if you can do something on your own, watching another person can help you hone your skills or pick up on details that might improve your behavior. Although the learning is faster, demonstrating the proper way to do something may result in a more superficial level of understanding compared to experimentation.

Modeling behavior not only applies to learning technical skills but soft skills as well, such as how to connect and interact with others. For example, bedside manner may not be something that new firefighters or EMTs should first learn through experimentation. It may be best for recruits to watch a role-play of veterans interacting with "patients" to copy body language and phrases that are acceptable for patient interaction. The phrase "leading by example" captures this form of behavior change.

Education does not require experimentation or observation of another person. Instead, it involves an exchange of ideas. For example, reading a book or taking a class causes an idea to be transmitted from the author or teacher to the reader or student. Education can involve introducing a new idea through writing, verbal communication, or a combination of the two. Education can be purely theoretical or skill specific.

Education runs the risk of the learner understanding the content in the shallowest way possible. Without the learner seeing or experiencing the subject matter, the leader relies on the learner's ability to concentrate on the message, imagine the correct message, process the meaning, and store the information. For hands-on skills that are dangerous or necessary for saving lives—like many skills in the fire service—this is not an ideal way to learn or relearn. On the other hand, education is the least restrictive form of transmitting knowledge because it is not bound by time or space. The teacher can be dead or alive or on the other side of the world, but the teacher is a leader if the learner voluntarily adopts the ideas in the pursuit of accomplishing a goal that improves the learner's body, mind, or soul.

***

It would be deficient as a leader to not use at least one of the three methods for reducing a knowledge gap. However, it can be excessive to try to do all of these methods all the time for each person or group. A good leader knows how to use each method and knows when to deploy them.

Deliver the learning plan

The second step of addressing a behavior change gap is to deliver the learning plan. Similar to the design phase, learning delivery can be an extensive topic and is beyond the scope of this book. You should consider three important factors when teaching someone new knowledge:

1. Formality
2. Choice of media
3. Proximity

Learning delivery can be either formal or informal. Most learning that occurs in the fire station is informal. Discussions around the dinner table or trainings in the engine bay occur during most shifts. The more complex the learning, the more consideration should be given to a formal learning approach.

Learning delivery can be done with any type of media or no media at all. Many effective low- and high-tech options for learning exist, including paper, computers, internet, radio, podcasts, television, and phones. If the learning is observational, there may be no need for media besides a leader's example. However, using some form of media allows for the learning to be documented and preserved, and thus it has the ability to be reviewed or reused in the future.

Learning delivery can be done in person or from afar. It can be done in the present moment or recorded for use in the future. Books and other forms of writing used to be the only way to transcend time and space, but with newer forms of media and the advent of cloud computing, many options exist that can allow for learning from different parts of the world and at different times in history.

Documenting targeted changes

The third step of addressing a behavior change gap is documentation. Documentation should include data for both the behavior that is being targeted for change and the targeted goal. Documentation increases accountability and allows for course corrections if the attempts at change fail. Documentation also provides evidence of both the change efforts and results so others can either adopt the change efforts if they were successful or examine the data in order to refute it. While documentation is not necessary, as complexity and time increase, documentation becomes more helpful.

Data on the targeted behavior

The targeted behavior refers to the new behavior a person is committed to adopting in order to change their reality. Collecting data on the targeted behavior helps an individual stay accountable for the behavior change. Since the focus is on the new behavior, the documentation does not require a comparison to the previous behavior. For example, if the person's goal is to lose 10 pounds (new reality), one targeted behavior might be to run for a minimum of 10 minutes each day. Depending on the desired level of data, documentation might include the person's heart rate while running, the mileage covered during the run, and how many minutes the run lasted. However, the documentation could be as simple as recording if the behavior took place or not. There will be times where it is deficient to not document anything, but documenting everything can be excessive. As a leader, you will have to find a balance.

Data on progress toward the targeted goal

The targeted goal is the new reality that someone is striving for. Collecting data on the individual's current reality will demonstrate whether any progress is being made. This would be done by comparing the current reality to the old reality. For example, for someone with a weight loss goal, daily or weekly weigh-ins would demonstrate if a person's efforts are working or not. If the efforts are not working, the goal or the learning plan may need to be changed.

Collecting these two types of data may not be necessary for all change efforts. For example, behavior change can occur with one day of skills training in which a firefighter learns how to more effectively throw a ladder. In the case of one-day skills enhancement training, documentation may be excessive. However, with long-term behavior change efforts, documentation may be the only way to know if the targeted behaviors have changed and if these changes are working toward the desired goal.

Motivation

Leadership is not about changing other people by force or mandates. Leadership is about supporting the evolution of the body, mind, or soul of others. That can be difficult to accomplish because each individual is at a different stage of willingness to change. You can be at one stage of willingness to change one thing but at a different stage of willingness to change something else. The Transtheoretical Model of Behavior Change (TTM) describes the different stages of behavior change.[11] It is important to be aware of these stages because if the person you are trying to lead is not ready, management science (including incentives and disincentives) may be the only way to change behavior.

The TTM describes the six stages of behavior change:

- **Precontemplation**: a person is not yet thinking of change
- **Contemplation**: a person is actively thinking about change
- **Preparation**: a person is preparing to make change
- **Action**: a person is working to change
- **Maintenance**: a person is working on keeping up the change
- **Termination**: a person has changed for good and no longer needs guidance

Based on TTM, a person has to be ready to take on their own evolution. For that reason, the TTM primarily focuses on motivation. Motivation (intention to choose) is the third area of Aristotle's assessment on responsibility. Sometimes, a person does not make the right choices because they lack knowledge of circumstances or knowledge of right and wrong. In those cases, reducing a knowledge gap will help lead a person toward the stages of action in the TTM. But what do you do if the person has the appropriate knowledge of circumstances and of right and wrong but continues to make the wrong choice? How do you motivate the person to voluntarily make the right choice?

Before we continue with motivation, it should be reiterated that DNA can account for someone's behavior that might appear to be choice but is perhaps genetically influenced.[12] For example, individuals with very low IQ may have a hard time learning new concepts or behaviors. Evidence shows that certain types of behavioral disorders are genetically influenced, such as depression, anxiety, and schizophrenia.[13] In extreme cases of genetic influence, this could mean the person is unlikely to change their behavior. It is worth stating explicitly that an awareness of genetic influences and differences should make you, as a potential leader, more caring toward others, not less.

Aristotle taught that everything has a purpose whether it is from nature or man-made. For example, the purpose of a firehose is to get water from the fire engine to the fire. The human purpose is happiness. However, happiness can be hedonic or eudaemonic. These are the two main ways that you can motivate others.

Hedonic motivation

Hedonic happiness comes from seeking pleasure, comfort, or enjoyment and from avoiding pain or discomfort. Hedonic motivation centers on the carrot (reward) or the stick (punishment). You can influence someone's decisions by promising them either one. You can even influence your own decisions by promising yourself either one. Reward or punishment applies physical, psychological, or spiritual pressure on your decisions, and therefore it limits your control over your own behavior.

Eudaemonic motivation

Eudaemonic happiness comes from the pursuit of challenging goals or individual fulfillment. You can motivate someone using eudaemonia by offering them the ability to thrive and fulfill both their purpose as a human (being sound in body, mind, and soul) and their individual purpose (reaching their goals). Offering someone the ability to fulfill their potential and to do things they value requires them to be in control of their own evolution. Ultimately, Aristotle argues that true happiness (eudaimonia) comes from living a virtuous life (avoiding deficiency and excess) in all that you do. This should not be mistaken with the idea of "trying everything in moderation." The idea of the virtuous life according to Aristotle was based in morality, which involved four cardinal virtues, among others. Those four cardinal virtues are prudence, justice, temperance, and courage.

∗∗∗

It is important to realize that both hedonic and eudaemonic motivation require knowledge. Doing something because it will help you to fulfill your potential requires knowledge of what your potentials are and how to reach them. Doing something because you will receive a reward or punishment comes from knowledge of rules or consequences. While both types of motivation may seem to be voluntary, hedonic motivation is the beginning of an influence or coercion campaign.

Management science asserts that at a certain point, most people will eventually act the way you want them to if offered enough reward or threatened with enough punishment. If leadership involved influencing others with reward or punishment, then leadership would increase with power, attractiveness, and money. As stated earlier, if that constituted as leadership, then there would be no use in trying to understand how to be a better leader; you would just need to get powerful, attractive, or rich.

Instead, leadership must be voluntary, which means the follower is free from force, influence, or coercion. That means that leadership must focus on closing the knowledge gaps as to what behavioral options are available to an individual or group; then, a leader should educate that person or group on how to pursue change for those behavioral options in the most virtuous ways possible, and they should bring awareness to the result of engaging in those options; then, a leader should let the person or group make their own determination about what to do. Assuming someone has knowledge of their circumstances and how to navigate an issue (knows right from wrong), then leading someone with eudaemonic motivation requires decreasing their knowledge gaps on why avoiding deficiency or excess is important for living a virtuous life.

There are many books on ways to increase motivation that go beyond the scope of this book. However, three eudaemonic motivators are worth considering, which were made famous in Daniel Pink's book, *Drive*.[14] If you are pursuing leadership as an action but you are not using eudaemonic motivators, then you are relying on hedonic motivators, and that is deficient. However, you do not have to use all of the eudaemonic motivators all of the time with everyone you are trying to motivate. As a leader, you will have to make decisions on when to use different motivators and who needs which one. The three main motivators offered by Pink are autonomy, mastery, and purpose.

Autonomy
Autonomy is the state of being self-governing. Autonomy increases motivation because it increases your sense of control. In other words, you trust someone to act with free will. As an example, if a person is told what to do but not how

to do it, they have more autonomy than someone who is told exactly what to do and how to do it.

To increase autonomy, you can increase a person's knowledge of what can and cannot be changed, as well as where they are free to be creative. Attribution theory proposes an explanation for motivation and behavior change that involves your sense of personal control.[15] The more you feel that your efforts caused an outcome, the more likely you are to contribute to that outcome. Therefore, increasing feelings of autonomy can be as simple as increasing someone's knowledge of the ways that their creative efforts can contribute to, and have an impact on, different outcomes and goals.

Mastery

Mastery is the possession or display of great skill or technique. Mastery increases motivation because you feel that you are capable of doing something with excellence. Mastery requires having someone believe that they can constantly improve, and that there is a path to improvement. Someone who feels like you cannot increase their mastery will be less motivated by you.

Self-efficacy theory proposes an explanation of motivation and behavior change that involves your sense of how well you can complete a specific action.[16] The more someone proves to themselves that they can handle a specific situation with excellence, the more likely they are to act in that specific situation. Therefore, increasing mastery can be as simple as increasing someone's knowledge on how to do a specific task.

Purpose

Purpose is an object or end to be attained. People want to feel that their skills are being used for something meaningful. Purpose involves having someone understand how their efforts are contributing to the organization, society, or even the world. Sometimes people can get so fixated on their task that they cannot see the forest for the trees. In other words, someone may need help seeing the bigger picture.

The utility theory of value proposes an explanation of motivation and behavior change that involves your perception of the usefulness of a task.[17] The more you feel that what you are doing is useful—either for the fulfillment of your goals or for the greater good—the more likely you are to do it. Therefore, increasing purpose can be as simple as increasing someone's knowledge of the value that their actions or contributions have on living a virtuous life or supporting their organization, society, or the world.

Part 3: Micro and Macro

Leadership and Organizational Change

An organization is a group of individuals who work together to achieve a goal. Achieving a goal is part of the definition of leadership. So, if an organization works to achieve a goal, which is part of the definition of leadership, does that mean the person in charge of an organization is a leader? According to the definition presented in this chapter, leadership requires positively changing the body, mind, or soul of humans by reducing a knowledge gap without the use of force, incentives, or disincentives. Executive decision-makers and middle managers can oversee the accomplishment of a goal. However, that does not necessarily mean they were leading, especially if they accomplished the goal by using management science techniques.

Take the example of a seasoned engine company of four with a lieutenant as the officer. A box alarm goes out and the engine company arrives on scene. The driver positions the engine and begins to set up water delivery. The lieutenant hops off and begins a 360°, and the firefighters begin to make entry. The fire is a room-and-contents on the first floor. The lieutenant makes it back around and catches up with the firefighters and they make a knock on the fire. The fire is out and the crew did an efficient job.

Did the lieutenant lead? Assume for a moment that the crew were not experienced and the lieutenant had to provide a number of commands, including where to position the engine, which lines to pull, and where to make entry. Does that constitute leading? If firefighters know what to do and will do it when they are told to do it, they are simply following orders. In fact, failure to do so could land them in trouble. It makes no difference if the incident were bigger and there were a chief giving orders.

Like humans, organizations have three dimensions, because organizations are just groups of individuals. If you take the body, mind, soul triad and abstract it to the organizational level, you end up with an action dimension (which is like the body), a business performance dimension (which is like the mind), and a collaboration dimension (which is like the soul). All businesses exist to take action on something in the world, they all have to manage their organization internally, and most organizations have to at least collaborate with the local government system but probably with other organizations as well. Additionally, each organization has to deal with people, processes, and things, all of which can be dealt with through management. Much like giving commands on a fireground, an executive decision-maker or manager can give directions in an organization. If the team or entire organization knows how to do their jobs and does what they are told to do, then they are simply following orders. No human development has taken place, and the following of orders was done because of rules, incentives, or disincentives. While the outcome is positive, it is still organizational management. It is easy for someone to achieve an organizational goal, especially if they have the ability to punish and reward. It is not easy to have a group of people voluntarily pursue and adopt positive changes to their body, mind, or soul by learning something new in order to improve an organization. That is organizational leadership.

The organization and you

> A chain is only as strong as its weakest link.

The links in a chain are often compared to members of a team. If one of the links fails, the chain fails. In the same way, team performance relies on the ability of each member to do their job. The relationship between a link and the chain is an oversimplification. A better comparison for the organization might be the body. Your body is made up of many cells, organs, and systems. It is possible for one cell to die or even become cancerous and still not cause death to the body. Even an organ or system that is dysfunctional in the body can be lived with, because the body has ways to overcome.

An organization is made up of many people, teams, and programs. Similar to the analogy of the body, a single person, or even a team or program, can be dysfunctional in an organization and the organization can still overcome and function. However, that says nothing about performance improvement within an organization. In order for an engine company to improve its performance, positive changes must occur in the body, mind, or soul of one or more firefighters in that company. In order for a battalion to improve its performance, positive changes must occur in one or more companies that make up the battalion. The same is true for the fire department as a whole. Figure 5–9 provides a model of how the individual firefighter connects to and impacts the fire department.

Group performance can only improve if individual performance improves. Individual performance changes might include improving strength or endurance (body), learning a new concept or skill (mind), or improving the connection or cohesion with the group (soul). Such individual performance changes can only occur through some form of learning which can be through experimentation, observation, or education.

In many organizations, physical fitness is not a requirement, and neither is team cohesion. Improving organizational performance is usually only a

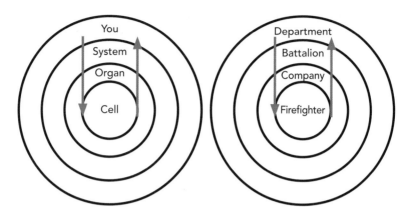

FIGURE 5-9. The group and you

question of individuals learning a new process or procedure. However, in the fire service, it is imperative that firefighters have the appropriate level of physical fitness, knowledge about the job, and connection to their team, because fulfilling the mission requires all three. The major goals of most fire departments are to do the job as efficiently, effectively, and safely as possible. In order to improve these areas, fire departments must improve their firefighters' bodies, minds, or souls. If nothing changes with individual firefighters, no improvements at the company, battalion, or organizational level can be expected.

An organizational leader is not a position. Rather, this title describes anyone who helps evolve individuals and groups in order to develop the organization, while using voluntary engagement and eudaemonic motivation—not rewards and punishment. If promotions are done correctly, a company officer should be able to develop the firefighters assigned to them, a station officer should be able to develop the company officers assigned to them, and the same is true for battalion chief, and so on. However, you do not have to be the fire chief to develop your organization. You can be an organizational leader if you possess knowledge that others do not and possess the ability to use that knowledge to help others change voluntarily. Therefore, organizational leadership is a matter of helping to evolve the bodies, minds, and souls of others by spreading your knowledge throughout the organization to create positive change. Rank is not required.

A competent fire chief should hypothetically be able to start a fire department from scratch by developing a few people into competent firefighters or officers. These firefighters or officers then develop others, and this cycle continues until an organization is formed. In this hypothetical scenario, the fire chief would have to create a system of communication and human interaction that develops people to fill certain roles that will keep the department running, which in turn continues to develop new fire officers and firefighters. The same is true with any organizational leader, except that the development can be anything that improves those within the organization toward a goal. Leadership at the organizational level includes the same necessary and sufficient components as leadership on an individual level. Leadership is not done by using control techniques or reward and punishment to get others to positively change. However, organizational leaders should understand how to develop and leverage systems to support the voluntary evolution of individuals.

Systems development

The human body has systems that develop and sustain the body automatically. To run these systems, the body requires only a minimal amount of resources (in the form of calories) based on the size of the body. Some of the systems include the immune system, central nervous system, and endocrine system. The systems complement and rely on each other. For example, the immune system monitors and responds to issues, the central nervous system learns and adopts new information, and the endocrine system controls growth and decay signals. The human body contains many more systems, all serving some purpose to keep the body in a healthy state. You cannot create new systems in the human body, but you can optimize the systems based on behavioral changes.

Organizations are similar to the human body. They have systems that develop and sustain the organization. However, the difference is that the systems in an organization are not developed or sustained automatically. Each one must be developed and maintained by people. In order to develop an organization, you either have to add new systems or optimize the existing ones. For systems that do not involve humans (e.g., finances or computer algorithms), only management is required. Management can also be applied to systems that involve humans, like the fire academy, if the intent is to control people's choices and behaviors. However, it is possible to create systems involving humans that do not control human choices or behavior, like optional career education. Systems also exist that support the voluntary interaction between people in order to develop them, such as peer fitness trainers or peer support teams.

As an organization grows, new systems will be necessary for optimal functioning. Consider the incident command system (ICS), which uses a generic organizational framework to help manage emergency responses. The ICS allows the incident commander to build out a response that is appropriate to the emergency, which can expand to become as big as necessary. The ICS provides the fundamental building blocks for an organizational system, but it is not a comprehensive guide to all of the systems an organization may need. As an incident grows, more components of an ICS are necessary. The same is true for organizational growth.

The organizational leader is one that realizes a need to introduce a new system or optimize an existing system within their organization. This is necessary as organizations grow and face new challenges. However, a leader rarely develops a completely new idea or system. Generally, a leader learns of an idea or system and brings it to an individual or group who is unaware of that idea. The fire service does this all the time. Smaller departments take ideas from larger departments. Additionally, national organizations exist that develop guidance and standards for fire departments (e.g., National Fallen Firefighters Foundation [NFFF]). An organizational leader in the fire service has many resources to pull from to bring change or development to their fire department. To introduce change into a fire department, the organizational leader should consider completing the same two steps for individual behavior change.

1. Assess the organizational behavior change gap: The leader would compare the organization's goals to where the organization is currently. Then, the leader can assess the organization using the Change Assessment Model. Finally, the leader can identify what learning is needed to address any knowledge gaps.
2. Address the organizational behavior change gap: The leader would create a learning plan that can be spread throughout the organization. To avoid using force, the system would have to spread the knowledge voluntarily. Additionally, data could be collected to document the change efforts and progress that is made.

These two steps can take years to accomplish at an organizational level. That is because it requires the organizational leader to: learn what is needed (if they do not already know); educate and train others who can then help in addressing the change gap; and leverage the appropriate resources, policies, and processes in order to allow the system to function. An organizational leader does not just change one individual. The organizational leader changes a group, or several groups, who will then help to support the evolution of the entire organization.

Developing a system rarely requires creating something new, and often only requires adopting and implementing systems that are already developed. For example, a new fire engine manufacturer uses components that already exist to create a new fire engine. Inventing something new is not a requirement to build a working fire engine. Organizational leadership is similar. A leader does not need to create new systems or ideas from scratch, but can simply use what already exists and structure it in a way that meets the organization's needs. In this way, you can think of organizational leadership as being a systems engineer for spreading behavior change.

However, organizational leadership consists of more than just leveraging or engineering systems. Organizational leadership refers to a larger-scale version of individual leadership. The point of leadership is to support individuals to evolve their body, mind, and soul to be more resilient so that they can overcome more challenges.

The same principle holds true for organizational leadership—it is about making your organization more resilient so that your organization can overcome more challenges. Recall from chapter 1 that the individual (micro) and the organization (macro) are similar to the cell and the body. What the organization does to help support the growth of individual firefighters will determine whether the organization as a whole grows more resilient (fig. 5–10).

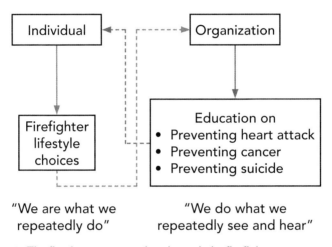

FIGURE 5–10. The fire department needs to invest in its firefighters.

A battalion chief should be continuously learning in order to pass on their knowledge and skills and develop station officers, who will then develop company officers, who will then develop firefighters. Similarly, an organizational leader of any rank should learn information and bring it to their department, then support the evolution of others who can use their new knowledge and experiences to help develop everyone else in the fire department. If the individual cells in our body don't grow, the body will not grow. If the people in your department don't grow, the department will not be able to overcome more challenges.

Health and wellness system: an example
Building a health and wellness system in your fire department requires positive change. It involves developing individuals, groups, and systems within your organization. To accomplish any or all of these goals requires a leader to "guide

on a way, especially by going in advance." The leader that sets out to develop a health and wellness system in their fire department should have advanced knowledge of health and wellness, how to lead individuals and groups, and how to lead organizations (which includes systems development). However, too often in the fire service, individuals are assigned to an administrative position based on rank or years of service as opposed to their competencies or credentials and their ability to help others evolve.

A health and wellness system is just a specific type of organizational system. There are three components that are critical for a system to be successful in your fire department. You need the support of your chief executive, a system design that fits your organization, and resources that allow the system to operate. In the fire service literature, publications are available that address each one of these components. To help you develop your own health and wellness system, below is a review of some of the most prominent publications on each component.

Chief executive support
The support of the chief executive is critical to getting things done within an organization. This is because the chief executive, or fire chief, is one of the key members of an organization who can set an organizational culture, which then determines the norms. One study found that the successful adoption of a fire department wellness system relied on two factors: a wellness champion and a supportive fire chief.[18]

The wellness champion could be a person at any rank in the fire department who operates as an organizational leader for wellness. The champion either has advanced competencies and credentials on health and wellness, leadership, and systems development or will set out to acquire those competencies and credentials and bring them to the department. The fire chief sets the tone for the department by modeling behavior of support for health and wellness to the other chiefs, officers, and department members. It is true that everyone in the department could reject the behavior modeled by the fire chief. However, without the support of the fire chief, any organization-wide change effort will likely be fruitless.

System design
Systems require a logical design in order to exist and continue to operate. A system provides the ability to support a certain process in a predictable manner. For example, the chain of command is a system that allows authority, responsibility, accountability, and communication to take place. The existence of a system might be enforced by rules or policy, but that does not necessarily mean that anyone is forced to engage with it. A system can exist to support

individuals in an organization only if they request support. Even though initial education and training may be required when hired, continuing education and training can be voluntary throughout a firefighter's career.

It is possible that a system intended to support and develop the body, mind, and soul of firefighters can be created and sustained by policy, while any firefighter's engagement with that system is entirely voluntary. In other words, a system is simply a tool that supports a process. The process can either be mandatory or voluntary. Some organizations provide guidance on what types of systems should exist for health and wellness in the fire service. Figure 5–11 provides a national-level summary of some of the most prominent guidance.

	Education	Wellness programs	Research
NFFF	Life Safety Initiatives #1, #5	Life Safety Initiatives #6, #13	Life Safety Initiative #7
IAFF/IAFC-WFI	WFI 4th edition, p. 8 (mission statement)	WFI 4th edition Ch. 3, 5	WFI 4th edition Ch. 7
NFPA	1500 Ch. 5 1583 Ch. 7, 8	1500 Ch. 4, 11, 12, 13 1583 Ch. 5, 6, 7, 8	1500 Ch. 4, 11 1583 Ch. 9
OSHA	Training requirements, pp. 1–4		
NIOSH		Kunadharaju, Smith, and Dejoy (2011) Hard et al. (2018)	

FIGURE 5–11. System design guidance from national-level organizations

The Occupational Safety and Health Administration (OSHA) requires that certain systems exist within organizations. Specifically, OSHA law states that employers are required to provide education to employees that aims to reduce the possibility of work-related injury, illness, or death. OSHA's mandate is general and affects all employers. State and municipal governments that regulate fire departments must meet or exceed OSHA's regulations. Additionally, the NFPA has included OSHA's regulation in its standards, as seen in chapter 4 of *NFPA 1500: Standard on Fire Department Occupational Safety, Health, and Wellness Program* and chapter 8 of *NFPA 1583: Standard on Health-Related Fitness Programs for Fire Department Members*. As firefighters demand more accountability from their fire departments on work-related health issues like heart disease, cancer, and behavioral health, fire departments should consider voluntarily building systems that can educate on those issues across an entire career.

Several national-level fire service organizations have published guidance on health and wellness systems. The NFFF published the 16 Firefighter Life Safety Initiatives which suggest that fire departments provide physical and behavioral support systems that are voluntary for members to engage in, and non-punitive. The International Association of Fire Fighters and International Association of Fire Chiefs Wellness-Fitness Initiative (IAFF/IAFC-WFI) has published guidance that suggests the same. The NFPA has published standards on health and wellness systems that fire departments should adopt when they are able to. Additionally, all three organizations suggest that fire departments should include education and research in their health and wellness systems. It should also be noted that the National Institute for Occupational Safety and Health (NIOSH), the organization responsible for line-of-duty death (LODD) investigations for firefighters, suggests that fire departments provide wellness support systems specifically because so many firefighter LODDs are caused by poor health.

You can use the material published from national organizations to guide your fire department on developing policies and rules that would support the creation and maintenance of a health and wellness system. Some aspects of a comprehensive health and wellness system might be mandatory, such as initial education or rehabilitative support for someone who is not capable of doing the job. However, most aspects of a health and wellness system can be voluntary, such as continuing education throughout a firefighter's career, in which someone who is seeking guidance on how to develop their body, mind, or soul can receive it.

Human resources
People, money, equipment, and information are a few examples of resources that may be necessary for systems to operate. However, people are the key component of systems that exist to educate and support the members of an organization. Published standards from the NFPA and guidance from the NFFF and IAFF/IAFC-WFI provide a framework for the human resources that are necessary to operate a health and wellness system.

> NFPA 1500 chapter 4 (Section 4.4.1) and NFPA 1583 (Section 4.1.1 and Section 8.1) are the standards for a fire department health and wellness system. The standards include information on physical health and fitness, behavioral health, and general wellness.

A fire department health and wellness system should provide preventative education resources as well as information on how to respond and treat injury or illness. The NFFF and IAFF/IAFC-WFI also provide guidance on fire department health and wellness systems. Both organizations have published documents that call for physical and behavioral health and wellness support systems, as well as encouraging spiritual support in the form of chaplaincy work. All three organizations recommend data collection and analysis.

A health and wellness system will look different in each fire department, but some structure or hierarchy may be necessary to manage it. Trained personnel will be necessary if the system requires data collection, analysis, and reporting. In addition, administrative personnel will be necessary if the system requires coordination and paperwork. You will also need knowledgeable trainers that can teach the support team members, who will then educate and support the rest of the members of the department. Because of unique needs, different people may be required for the physical, behavioral, and spiritual support.

A wellness champion might serve as a health and wellness coordinator who manages all aspects of the health and wellness system. If the wellness champion attempts to act as the organizational leader of health and wellness in their fire department, even if some managerial duties are necessary to build or maintain a system, the positive development in others has to be voluntary; otherwise, it is not leadership.

The IAFF/IAFC-WFI has published a model of how support teams can be organized. As a special assignment, teams of firefighters could volunteer to become peer fitness trainers or peer support team members who would serve as field level specialists for physical and behavioral health respectively. These peer specialists would be the members in a department that would respond to physical or behavioral assistance requests, but would also proactively work to educate members on how to develop themselves through voluntary engagement. The organizational leader for health and wellness would work to develop these peers, who would then work to develop members of the department, which then improves the whole department.

In addition to peer support, more advanced human resources may be necessary to support a health and wellness system. For instance, a chaplain could coordinate and provide spiritual support for firefighters from all faiths. A chaplain can work in cooperation with the behavioral support system or work alone. Mental health professionals and physical therapists could provide higher-level support that goes beyond the scope of practice of specialty team members in the health and wellness system. Additionally, these more advanced human resources can help provide continuing education for the wellness champion and the members of the health and wellness system.

Members of the health and wellness system should be able to provide initial education, continuing education, and remedial education as needed. They should operate whenever possible under voluntary engagement with the members whom they are working to develop. These are the people who will lead the charge on improving the body, mind, and soul of firefighters.

Conclusion

1. Leadership and management are similar but are not the same
2. Leadership can be objectively assessed when we identify the necessary and sufficient components
3. Leadership as a verb (action) does not rely on rank or position
4. Organizational leadership is simply a scaled-up version of leadership

Leadership is hard. It requires thoughtfulness, precision, and purposefulness. Leadership effects a voluntary, lasting, positive change to the body, mind, or soul of an individual or group toward an established goal by reducing a knowledge gap. Ultimately, leadership is about helping individuals or groups evolve their body, mind, and soul.

Good leaders need to know how to manage when necessary, because not everyone is ready or willing to change. Ideally, all managers would know how to effectively lead, but that is not always the case. Management is a science based in psychology, because well-defined principles of human behavior exist that can be controlled. Organizations need management because organizations are held accountable for their actions by stakeholders, and they need to ensure employees are doing their jobs. Perhaps it could be said for organizations that management is what we rely on when leadership is not working.

Leadership is rare. Otherwise, everyone would be a leader and there would be no need or desire for people to learn about leadership. In fact, the opposite is true. People instinctively know that not everyone is a leader and that leaders are special. Leadership requires certain abilities like critical thinking, planning, and communicating. In that sense, not everyone can be a leader even if they try. However, even though not everyone is a leader and not everyone can be a leader, those that have the ability and the drive can learn to be a leader.

The writer Will Durant poignantly paraphrased Aristotle when he wrote, "We are what we repeatedly do." This statement means that your reality is determined by your behaviors and your habits, which rely on the knowledge you are given by your parents, your communities, and the leaders that guide you. Aristotle taught that people can improve. But people do not improve in a vacuum. People are improved by learning how to be a better version of themselves. Leadership is about supporting those improvements voluntarily. Yes, we are what we repeatedly do, but we do what we repeatedly see and hear.

Questions

1. What are the necessary and sufficient components of leadership?
2. What are the three dimensions of a human?
3. Why does leadership require change toward a goal?
4. Why must change from leadership be positive?
5. What is the difference between change being objectively or subjectively positive?
6. Why is reducing a knowledge gap required for leadership?
7. Why does leadership have to be voluntary?
8. What does the use of incentives or disincentives by someone in charge demonstrate?
9. Why does any change have to last in the leader's absence?
10. What is the shortened definition of leadership?
11. What are three things you can change as a leader?
12. What are the three steps to assessing the behavior change gap?
13. What are the three steps to addressing the behavior change gap?
14. What are the three different ways to reduce a knowledge gap?
15. What is the Trans-Theoretical Model (TTM) of Change?
16. What are the two different ways that you can motivate someone?
17. What three motivational tactics did David Pink make famous?
18. What three things are critical for a system to be successful in your department?
19. Which chapter in NFPA 1500 addresses fire department health and wellness systems?
20. Which chapters in NFPA 1583 address fire department health and wellness systems?

Notes

1. Merriam-Webster, s.v. "lead (v.)," accessed July 24, 2023, https://www.merriam-webster.com/dictionary/lead.
2. Merriam-Webster, s.v. "manage (v.)," accessed July 24, 2023, https://www.merriam-webster.com/dictionary/manage.

3. Vineet Nayar, "Three Differences Between Managers and Leaders," *Harvard Business Review*, August 2, 2013, https://hbr.org/2013/08/tests-of-a-leadership-transiti.
4. Merriam-Webster, s.v. "human (n.)," accessed July 24, 2023, https://www.merriam-webster.com/dictionary/human.
5. Merriam-Webster, s.v. "health (n.)," accessed July 24, 2023, https://www.merriam-webster.com/dictionary/health.
6. George L. Engel, "The Need for a New Medical Model: A Challenge for Biomedicine," *Science* 196, no. 4286 (1977): 129–136, https://doi.org/10.1126/science.847460.
7. Aristotle, *Nicomachean Ethics*, trans. W. D. Ross (2005), http://classics.mit.edu/Aristotle/nicomachaen.html.
8. Merriam-Webster, s.v. "hedonic (adj.)," accessed July 24, 2023, http://www.merriam-webster.com/dictionary/hedonic.
9. Merriam-Webster, s.v. "eudaemonic (adj.)," accessed July 24, 2023, http://www.merriam-webster.com/dictionary/eudaemonic.
10. Loren L. Toussaint, E. L. J. Worthington, and David R. Williams, *Forgiveness and Health* (Springer, 2015), https://doi.org/10.1007/978-94-017-9993-5.
11. James O. Prochaska and Wayne F. Velicer, "The Transtheoretical Model of Health Behavior Change," *American Journal of Health Promotion* 12, no. 1 (1997): 38–48, https://doi.org/10.4278/0890-1171-12.1.38.
12. Robert Plomin et al., "Top 10 Replicated Findings from Behavioral Genetics," *Perspectives on Psychological Science* 11, no. 1 (2016): 3–23, https://doi.org/10.1177%2F1745691615617439; Laura A. Baker, "The Biology of Relationships: What Behavioral Genetics Tells Us About Interactions Among Family Members," *De Paul Law Review* 56, no. 3 (Spring 2007): 837.
13. E. Pettersson et al., "Genetic Influences on Eight Psychiatric Disorders Based on Family Data of 4,408,646 Full And Half-Siblings, And Genetic Data of 333,748 Cases And Controls," *Psychological Medicine* 49, no. 7 (2019): 1166–1173, https://doi.org/10.1017/s0033291718002039; Robert Plomin, *Blueprint: How DNA Makes Us Who We Are* (MIT Press, 2019).
14. Daniel H. Pink, *Drive: The Surprising Truth About What Motivates Us* (Penguin, 2011).
15. Eric Anderman and Lynley Anderman, "Attribution Theory" (The Gale Group, 2006), http://hillkm.com/yahoo_site_admin/assets/docs/anderman_e_anderman_l_2006_attributions_education.com.pdf.
16. Frank Pajares, "Self-Efficacy during Childhood and Adolescence: Implications for Teachers and Parents," in *Self-Efficacy Beliefs of Adolescents*, ed. Frank Pajares and Timothy C. Urdan (Greenwich, CT: Information Age Publishing, 2006), 339–67.
17. Allan Wigfield and Jacquelynne S. Eccles, "Expectancy–Value Theory of Achievement Motivation," *Contemporary Educational Psychology* 25, no. 1 (2000): 68–81, https://doi.org/10.1006/ceps.1999.1015.

18. Hannah Kuehl et al., "Factors in Adoption of a Fire Department Wellness Program: Champ and Chief Model," *Journal of Occupational and Environmental Medicine* 55, no. 4 (2013): 424, https://doi.org/10.1097%2FJOM.0b013e31827dba3f.

ANSWERS TO END OF CHAPTER QUESTIONS

CHAPTER 1

1. "We are what we repeatedly do. Excellence, then, is not an act, but a habit."
2. "We are what we repeatedly do" describes that, as people, our present condition is the result of our actions. "Excellence, then, is not an act, but a habit" elucidates that we don't become excellent from one act; rather, it is from repeatedly choosing the right behaviors over time.
3. Health is not decided from one decision or action. It comes from a pattern of behavior repeated over time.
4. Merriam-Webster defines health as being sound in body, mind, or spirit.
5. Wellness is what determines the health of your component parts as a human. Wellness is the input, and health is the outcome.
6. Resilience is a measurement of how much challenge your body, mind, and soul can overcome.
7. FIRST USE (the ability to bounce back): Humans naturally change with age, but they can also change through wellness efforts. Humans have a body, mind, and soul (or bio-psycho-social dimensions, if you prefer) and thus bouncing back would have to relate to more than just your physical body. Calling a person resilient in a situation that may lead to death, when there are better options to help you survive, misses the point.
SECOND USE (having more in the tank): If reserves are what make humans resilient, then more is simply better. In other words, being as muscular as possible, as fat as possible, or as rich as possible should make you more resilient because you would always have more reserves. But this is not the case.

8. If you compare two people with the same baselines, resilience is determined by conditioning, or the effort someone puts in to grow their abilities. All things being equal, the person who has better physical conditioning will be able to overcome more physical challenges. Similarly, with the same IQ as a baseline, the person who learns and practices a skill more will be able to perform better than the person who does not. For those that believe love and forgiveness are abilities of the soul, the one who practices loving and forgiving others is better conditioned in those areas.
9. Heart attack, cancer, and suicide.
10. Your wellness habits influence your physical and behavioral health, and poor physical and behavioral health issues that are ignored or unaddressed increase the likelihood of the Big Three fatalities.
11. NFPA 1500 and 1583.
12. NFPA 1500: Chapter 11 covers medical and physical requirements; chapter 12 covers behavioral health and wellness programs.
13. NFPA 1583: Chapter 6 covers fitness assessments; chapter 7 covers exercise and fitness training programs; chapter 8 covers health promotion education.
14. The Occupational Safety and Health Administration (OSHA) mandates that employers train their employees on how to reduce or eliminate work-related illness, injury, and death.
15. The Public Safety Officers Benefit Act (PSOB) is a federal law that allocates tax dollars for families of fallen public safety officers who have died in the line of duty.
16. Currently, the PSOB includes heart attack and stroke, which can each be the direct result of lifestyle factors. Therefore, firefighters should at least be taught how to prevent heart attack and stroke, since those are considered job-related, line-of-duty deaths.
17. 1. Did the individual know the circumstances?
    2. Did the individual know right from wrong?
    3. Did the individual voluntarily choose their action?
    4. Did the individual have the ability to do otherwise?
18. Figure 1–6 graphically presents Aristotle's four assessments using three pairings. Nature and nurture represent the circumstances of what you cannot change (nature) and what you can change (nurture). The spectrum of deficiency and excess exists for both knowledge of right and wrong as well as choice. Assessing the relationship between micro and macro allows you to identify the forces that may influence your choice.

19. If it is true that firefighters are more stressed than the average worker, then in order to work toward the virtuous middle, firefighters should put more effort into wellness than the average worker. However, the research suggests that not only are firefighters more stressed, a significant number have worse wellness habits than the average worker.
20. A 90% preventability rate is a consistent theme across heart attack, cancer, and suicide.

CHAPTER 2

1. You have certain biological features that you have to live with. Three of those features are the stress response system, the inevitable decline of naturally occurring human growth hormone (HGH), and your genetically inherited body type.
2. Exercise, fasting, and sleep, all of which can optimize your HGH naturally.
3. Here is the secret about exercise: it does not matter which exercise you do, as long as you are exercising enough days per week (frequency), for enough time (duration), and with enough effort (intensity). Each person is different, so "enough" will depend on your needs. Those three factors represent exercise volume ($V = F \times D \times I$).
4. The measurement of 1 MET is your exertion level while at rest (e.g., sitting).
5. Vigorous activity is anything above 6 METs and includes jogging (roughly 6 METs), body weight exercise (roughly 8 METs), and jumping rope (roughly 10 METs).
6. According to NFPA 1582, firefighters need to be able to hit 12 METs on a stress test (Bruce protocol) for their annual physical medical evaluation.
7. HGH is secreted only after you enter Gear 2. It takes roughly 10 minutes in Gear 2 to begin producing HGH.
8. Intermittent fasting falls into two main categories. The first category is time-restricted feeding (TRF), which allows you to refuel each day during specific windows of time. The second category is full-day fasting, which involves a 24-hour fast.
9. There are three common models of TRF, starting with fasting periods of 12 hours and becoming more restrictive: 12 hour +, 16:8, and

Warrior. Full-day fasting involves 24 hours (or more) of not eating and includes periodic fasting and alternate day fasting.
10. The 12-hour threshold for HGH is similar to the threshold between Gear 1 and Gear 2 for exercise. If you do not cross the threshold, you will not get HGH from fasting.
11. Throughout the night, your brain attempts to complete natural sleep cycles which take you from a conscious state into stage 1, then stage 2, then stage 3, and then to REM sleep.
12. In stage 3, your body secretes HGH.
13. The suggested amount of sleep for the average person is 6–8 hours per night. In figure 2–16, you can see 6–8 hours of sleep each night means 42–56 hours of sleep per week.
14. The concept of the S-curve is that change occurs at different speeds on the road to mastery. Initially, change is slow until you cross a certain threshold. After that, rapid growth becomes possible and the payoff of effort is greater. Eventually, growth slows down and becomes incremental, and the payoff of effort becomes much less.
15. HEART ATTACK: Compared to non-runners, those who ran had a 45% lower risk of cardiac mortality. CANCER: Healthy mitochondria can actually suppress the reproduction of cancerous cells. As it turns out, exercise plays a significant role in the health of your mitochondria. Exercise not only increases the ability of mitochondria to function; it also increases the number of mitochondria you have in your cells. SUICIDE: Poor blood flow to the brain is also associated with depression, ADHD, bipolar disorder, addiction, schizophrenia, and suicide.
16. HEART ATTACK: Fasting has been shown to optimize body systems that affect the heart, including lipid metabolism (cholesterol balancing), reduction of inflammation, reduction of body weight, reduction of blood pressure, reduction of resting heart rate, and optimization of glucose and insulin. CANCER: When an organism goes into a negative energy balance, a trigger causes the organism to find cells that are dying, weak, or even cancerous and devour those cells first for energy metabolism. SUICIDE: Fasting can stimulate neurogenesis (the growth of new brain cells) by producing more of the chemical called brain-derived neurotropic factors (BDNF). BDNF allows for new neural connections to be made, repairing ageing brain cells and protecting healthy cells, and is now believed to play a role in stress resistance and resilience.
17. HEART ATTACK: A meta-analysis of 11 studies that included over 1 million people found that those who slept less than six hours or

more than eight hours, on average, had a higher risk of dying from coronary artery disease or stroke. CANCER: In a 2012 study that spanned six years with over 3,100 men, researchers found that, compared with men that never worked night shift, those who did had an increased risk of cancers including lung, colon, bladder, prostate, rectal, and pancreatic cancer and non-Hodgkin's lymphoma. SUICIDE: Researchers found that the more change there was in the time that a participant went to bed and woke up, the higher the suicide ideation risk. The results were the same even when the researchers controlled for depression, substance use, and suicidal symptoms at the start of the study.
18. Chapter 4 (section 4.3), chapter 11 (sections 11.2, 11.3, and 11.7), and chapter 12 (section 12.2).
19. Chapter 5, chapter 6, chapter 7, and chapter 8.
20. NFPA 1500 chapter 4 (section 4.3.1), chapter 5 (section 5.1.1); NFPA 1583 chapter 8 (section 8.1.1).

CHAPTER 3

1. There are certain nutritional principles that cannot be changed because they are part of human biology. Three of these principles are nutrient requirements, nutrient function, and nutrient impact.
2. There are three factors of nutrition that you can control. You can control your nutrient variety, nutrient density, and nutrient quantity.
3. You cannot change the fact that the human body requires certain materials in order to sustain itself and operate the systems that allow the body to function. Those materials come in two basic categories: caloric or non-caloric.
4. In the MyPlate model, you see the five major food groups—fruits, vegetables, grains, protein, and dairy—represented on the plate. In general, the advice is to aim for ½ of your plate to be fruit or vegetables, ¼ of your plate as protein, and ¼ of your plate as grains. A serving of dairy is depicted as additional to the plate.
5. It is important to choose a variety of colors because each color provides different nutrients in your diet.
6. Nutrient Density = Nutrients/Calories
7. Some of the different ways that food is processed include: mechanical, chemical, and state.

8. The soil quality will impact the quality of fruit, vegetables, and grains; the feed quality will impact the quality of meats, fish, and eggs; fruit, vegetables, and grains may be heavily sprayed with chemicals; meats and fish may have been raised in a habitat that is inhumane or unnatural; meats and fish may have been injected or fed antibiotics, hormones, and other chemicals; the fruit, vegetables, and grains may be genetically modified (GMO); meats and fish may be GMOs or may have been fed GMO products.
9. The USDA regulates organic labeling. "Organic" means that animals were raised without antibiotics or hormones and cannot be fed GMO products. Plants are produced without conventional pesticides, synthetic fertilizers, sewage sludge, bioengineering, or ionizing radiation. In order to qualify for the USDA Organic seal on a food product, the product must have at least 95% organic ingredients.
10. Glyphosate is the main ingredient in Roundup. In 2015, the International Agency for Research on Cancer (IARC) classified glyphosate as a "probable carcinogen" to humans. There have been many lawsuits against Monsanto (the original owner) and Bayer (the new owner), including an $11 billion settlement that covers roughly 100,000 claims.
11. The "Dirty Dozen" list is published annually to warn consumers about which non-organic fresh foods have the highest levels of chemicals. The list can be found on EWG's website at www.EWG.org.
12. The Glycemic Index is a standardized system (from 0–100) that rates foods on how much impact the food has on your blood sugar levels compared to pure glucose. The University of Sydney (Australia) maintains a Glycemic Index database. The database is available to anyone with internet access at www.glycemicindex.com.
13. The Aggregate Nutrient Density Index is a method of rating the nutrient density of food and was created by Dr. Joel Fuhrman. Dr. Fuhrman argues that the nutrient density in your body comes from the nutrient density in your food, and therefore your diet is a major factor in whether you will suffer from major lifestyle diseases.
14. The Eat to Beat Cancer campaign is an initiative to help fight the cancer epidemic by eating foods that have cancer-fighting properties. Dr. William Li is the CEO and Medical Director of the Angiogenesis Foundation. The foundation's team has developed a database of natural cancer-fighting foods with the evidence to back it up. The database can be found at www.eattobeat.org/food.
15. Easy Caloric Need Estimate: Weight × 15

16. Easy Hydration Estimate: ½ your body weight in ounces of water.
17. To lose 1 pound, you have to decrease your calories by about 3,500 per week.
18. HEART ATTACK: The 7 Countries Study compared diets from U.S., Finland, Netherlands, Italy, Greece, Croatia, Serbia, and Japan. The study found that the food habits of the Mediterranean countries led to a 39% decrease in mortality from heart issues and a 29% decrease in mortality from blood vessel issues compared to the other regions studied. CANCER: The largest contributor of cancer from lifestyle factors was diet, making up 30%–35%. Tobacco made up 25%–30%, infections made up 15%–20%, obesity made up 10%–20%, and alcohol made up 4%–6%. The remainder was compiled into a group called "other." This research suggests that nutritional behavior (diet, obesity, and alcohol) plays a role in up to 61% of all cancers. The study explains that diet specifically is linked in up to 70% of colorectal cancers. SUICIDE: A meta-analysis involving many different dietary patterns found that those who adhere to a Mediterranean diet had a 33% lower risk of depression.
19. NFPA 1500 Annex-A Explanatory Material (Section A.12.2.1).
20. NFPA 1583 Annex-A Explanatory Material (Section A.8.1.2).

CHAPTER 4

1. The World Health Organization (WHO) defines mental health as "a state of well-being in which the individual realizes his or her abilities, can cope with normal stresses of life, can work productively and fruitfully, and is able to make a contribution to his or her community." The Centers for Disease Control and Prevention (CDC) describes mental health as an individual's "emotional, psychological, and social well-being."
2. DISTRESS: Psychological distress is a general term describing a variety of negative psychological symptoms like excessive worry, intense pressure, and emotional strain. It is an adverse response to stressors in your life. WELL-BEING: Psychological well-being is a positive mental state that has two components. The first is called hedonic well-being, which refers to your subjective feelings of happiness. The second component is called eudaemonic well-being, which refers to optimal mental wellness efforts. It has six

components—self-acceptance, environmental mastery, positive relationships, personal growth, identifying your purpose in life, and autonomy.
3. BRAIN: The brain is an organ protected by the skull. It is the epicenter of the nervous system, a collection of billions of interconnected neurons that control and regulate behavior. MIND: The mind, however, is harder to explain because it is not a physical thing. Rather, it is a word that symbolizes and describes all mental processes like thinking, feeling, remembering, learning, and judging.
4. Principle 1: All mental processes reflect brain processes. Principle 2: Genes (and their combinations) control brain functioning, which in turn controls behavior, including mental functions and mental illness. Principle 3: Social, developmental, and environmental factors alter gene expression. Principle 4: Alterations in gene expression induce changes in brain functioning. Principle 5: Learning changes behavior by producing changes in gene expression that alter the strength and structure of the brain.
5. The degree to which people struggle or succeed in changing their behavior is partially influenced by genetics, particularly in two areas—intelligence and sensory function.
6. There are two major behaviors under your control—your thought patterns and your social connection habits.
7. Limbic resonance is the activity in your brain that occurs when you share a meaningful connection with someone. This happens because your limbic systems actually connect with each other. Being seen and heard by another person helps regulate your brain activity. As a result, you develop stronger connections with that person.
8. DOING: Doing represents purposeful, goal-orientated activities like remodeling your kitchen or running a marathon. Regardless of the activity, doing has structure—it is the means to an end. But it also builds competence, enhancing your feelings of self-worth. BEING: Alternatively, being activities are unstructured and contemplative. Being includes reflection, meditation, self-(re)discovery, and savoring the moment. Being is not a means to an end—it's listening to music or going on a walk.
9. To provide some clarity, Dr. Shauna Shapiro has a clear and concise definition of "mindfulness": intentionally paying attention in a kind, open way.
10. While the definition may sound vague, it is built around three core ideas—attention, intention, and attitude. As described in more detail below, these core ideas show how mindfulness is a type of lived

mediation—a user-friendly way of taking control of your mind so you can focus without getting distracted.
11. GUIDELINE 1: Imagine yourself doing the task with as much detail as possible. GUIDELINE 2: Imagine failing as well as succeeding, because failure is going to happen. GUIDELINE 3: Group input can enhance the effectiveness of mental rehearsal. An overlooked type of mental rehearsal is the after-action review (AAR). Originally developed for the U.S. Army, the AAR requires both self and group reflection.
12. Strengthen connections; be intentional with technology; develop helpful routines and habits.
13. EMPATHY: Empathy is the glue required to connect with people and build relationships. The Oxford Learner's Dictionary defines empathy as "the ability to understand another person's feelings or experiences." SELF-COMPASSION: Popularized by Dr. Kristin Neff, a professor of psychology, self-compassion is kindness shown to yourself when you suffer, fail, or feel inadequate. It is protection from acidic thoughts or emotions caused by stressful situations.
14. SIMULATION THEORY: Simulation theory suggests that empathy is an automatic process caused by mirror neurons. MIND THEORY: Mind theory posits that empathy is not an automatic process. Rather, it is a focused cognitive process.
15. Change your perspective; supportive self-talk; SWOT analysis.
16. The model requires balancing the two pairs of opposing needs described earlier in this section: doing versus being and solitude versus relationships. As seen in figure 4–7, combining these two pairs creates four categories: solitary doing, communal doing, solitary being, and communal being.
17. Neuroplasticity is the brain's ability to change and adapt because of experience, allowing you to learn new things, unlearn old things, improve existing cognitive abilities like memory or attention, and even recover from traumatic brain injuries.
18. ALZHEIMER'S: Although much is still unknown, it appears that managing stress and depressive symptoms helps to prevent the cognitive declines associated with Alzheimer's. SUICIDE: Social isolation and loneliness are important risk factors associated with suicidal thoughts and behaviors. As a result, strengthening relationships may help prevent suicidal tendencies. CANCER: In large meta-analysis studies, it appears that stress is linked to the growth of cancer and cellular aging. The main reason seems to be that chronic stress increases the inflammatory response in your body

while also lowering your immune surveillance, which then leads to the development and growth of tumors. HEART ATTACK: The INTERHEART study, which took data from 52 countries, found that chronic stress increases your chances of having a heart attack by over 200%!
19. NFPA 1500 chapters 12 and 13. Additional information can be found in NFPA 1500 Annex A (Section A.12 and Section A.13).
20. NFPA 1583 chapter 8 (Section 8.1.4).

CHAPTER 5

1. Leadership must involve humans; leadership must involve change toward a goal; the change must be positive; the change reduces a knowledge gap; the change must be voluntary; the change must last in the absence of the leader.
2. Body; mind; soul. Alternatively, biological; psychological; sociological.
3. Leadership is not about change for the sake of change alone. Change without a goal can happen through exploration or experience, but those instances do not require a leader. It is also problematic to suggest that someone follow another person without a goal that appeals to the follower.
4. Leaders do not make neutral or negative changes. Neutral and negative changes to an individual or group provide no benefit from baseline. Furthermore, any negative change to an individual or group would require involuntary action or the use of incentives, because no sane individual or group would voluntarily negatively impact an aspect of their body, mind, or soul without believing there was a net benefit to one or more of their dimensions. Organizations do not train people to become leaders in hopes that they may make neutral or negative changes. Leaders make positive changes. Leaders must improve individuals and at least not harm society.
5. It is important to note that assessing change can be objective or subjective. An objective assessment of change considers a new status compared to a baseline status. A subjective assessment of change offers a judgement on the value of change. Change can be objectively positive, but subjectively negative.

6. Someone who knowingly provides false information to another is attempting to deceive that person, and that is not leadership. Someone who mistakenly or ignorantly provides false information to another is not leading because they don't actually understand what is happening. In both cases, an actual knowledge gap was not closed because the information was false. Without knowledge to pass on to someone else, the most you can say about a person who was trying to lead another person or group is that they were co-exploring. Without the passing of knowledge, the leader cannot expect any change to occur in the follower and therefore the leader-follower relationship is not necessary.
7. Aristotle taught that without voluntary choice, an individual cannot be fully responsible for their actions. Part of voluntary choice involves the ability to do otherwise (choose differently). Management science relies on reducing the freedom of choice by controlling the environment or resources or by using enough incentives or disincentives to control the choices of individuals and groups. It is true that you always have a choice, even if the choice involves your own death. However, when doing leadership, the threshold for freedom of choice ends the moment any control factors are used, no matter how small.
8. The use of incentives or disincentives by someone in charge demonstrates their belief that they would not be followed otherwise.
9. It is one thing for a person to voluntarily learn something new from someone else. It is another thing for a person to voluntarily use that knowledge to change their behavior while the person who taught them the new knowledge is around. It is yet another thing entirely for a person to adopt that behavior change when the leader is absent. The longer the change lasts in the absence of the leader, the more we can conclude that the change was adopted voluntarily and valued by the follower; if there is no active follower who has adopted the guidance of the leader and continues the behavior on their own, then leadership is not occurring.
10. The purpose of leadership is to help humans be more capable of overcoming challenges—that is, to be more resilient. At its core, then, *leadership is supporting the evolution of another person's body, mind, or soul.*
11. There are three areas of leadership that you can change. You can change which dimension of the human you focus on—body, mind, or soul. You can also change the way in which you close the knowledge

gap. Lastly, you can change the way you motivate someone to voluntarily engage in the evolution of their behavior.
12. Assessing behavior change includes identifying a goal, establishing a baseline, and identifying the required learning needed to close the gap.
13. Addressing the behavior change gap is a matter of successfully developing a learning plan, delivering the learning plan, and documenting the targeted changes.
14. Experimentation; observation; education.
15. The Trans-Theoretical Model of Change (TTM) describes the different stages of behavior change. It is important to be aware of these stages, because if the person you are trying to lead is not ready, management science (including incentives and disincentives) may be the only way to change behavior. The six states of the TTM are pre-contemplation; contemplation; preparation; action; maintenance; termination.
16. HEDONIC MOTIVATION: Hedonic happiness comes from seeking pleasure, comfort, or enjoyment, and avoiding pain or discomfort. Hedonic motivation centers on the carrot (reward) or the stick (punishment). You can influence someone's decisions by promising them either one. You can even influence your own decisions by promising yourself either one. Reward or punishment applies physical, psychological, or spiritual pressure on your decisions and therefore it limits your control over your own behavior. EUDAEMONIC MOTIVATION: Eudaemonic happiness comes from the pursuit of challenging goals or individual fulfillment. You can motivate someone using eudaimonia by offering them the ability to thrive and fulfill both their purpose as a human (being sound in body, mind, and soul) and their individual purpose (reaching their goals). Offering the ability to fulfill someone's potential and to do things that they value requires them to be in control of their own evolution. Ultimately, Aristotle says that true happiness (eudaimonia) comes from living a virtuous life (avoiding deficiency and excess) in all that you do. This should not be mistaken with the idea of "try everything in moderation." The idea of the virtuous life according to Aristotle was based in morality, which involved four cardinal virtues, among others. Those four virtues are: prudence, justice, temperance, and courage.
17. The three main motivators offered by Pink are autonomy, mastery, and purpose.

18. There are three components that are critical for a system to be successful in your fire department. You need the support of your chief executive, a system design that fits your organization, and resources that allow the system to operate.
19. NFPA 1500 chapter 4 (Section 4.4.1).
20. NFPA 1583 chapters 4 and 8 (Section 4.1.1 and Section 8.1).

INDEX

A

AAR (After Action Review) 151
ACE. *See* American Council on Exercise (ACE)
active state 38
ADF (alternate day fasting) 53
ADHD (attention deficit hyperactivity disorder) 66, 138
adrenaline 142
AFG (Assistance to Firefighters Grants) 25
After Action Review (AAR) 151
Aggregate Nutrient Density Index (ANDI) 102
aging 32
alcohol 86
    esophageal cancer and 91
    intolerance to 91
alternate day fasting (ADF) 53
Alzheimer's disease 169
American Council on Exercise (ACE) 36
    Health Coach Manual 109
American Psychological Association (APA) 70
amygdala 143, 168
ANDI (Aggregate Nutrient Density Index) 102
angiogenesis 87, 115
    cancer and 102
Angiogenesis Foundation 102
antibiotics 95
antidepressants 68, 171
antioxidants 86

anxiety 138, 211
    chronic diseases and 142
    cortisol and 142
    depression and 143
    fear network and 142
    suicide and 116
APA (American Psychological Association) 70
Aristotle, on
    assessing behavior 199–202
    health and wellness 2
    human purpose 211
    individual autonomy 189
    individual differences 60
    individual responsibility 203
    responsibility assessment 12–14
    the virtuous life 211
Assistance to Firefighters Grants (AFG) 25
attention and mindfulness 148
attention deficit hyperactivity disorder (ADHD) 66, 138
attitude and mindfulness 149
attribution theory 213
autism 138
autonomy 212
autophagy 55, 67. *See also* fasting

B

Bandura, Albert 22
baseline measurement 204–205
BDNF. *See* brain-derived neurotropic factor (BDNF)

behavior 23
    assessment of 199–202
Behavioral Risk Factor Surveillance
    System (BRFSS) 113
behavior change 145, 193–194, 203–205
    baseline measurement of 204–205
    data and 210
    goal-setting and 203
    stages of 210–211
    Transtheoretical Model of Behavior
        Change (TTM) 210
big three 8, 16, 169–170. *See also* cancer;
    heart attacks; suicide
    nutrients and 86
    risk reduction of 20
biopsychosocial dimensions 4
biopsychosocial triad 183–184
blood sugar 102. *See also* Glycemic
    Index (GI)
BMI. *See* body mass index (BMI)
body 198–199
body fat 36–37, 40, 66, 86, 204. *See
    also* body mass index (BMI);
    Human Growth Hormone
    (HGH)
    brain function and 116
body mass index (BMI) 36–37
    measuring 36
body-mind-soul triad 183–184
body systems 217
body types 34–36
    ectomorph 35–37
    endomorph 35–36
    mesomorph 35–36
body weight 67
bouncing back 3–4
brain-derived neurotropic factor
    (BDNF) 66, 68
    exercise and 147
brain health 66
    body fat and 116
    gray matter and 116
brain-mind framework 134–137, 198
    behaviors and 135
    external factors and 136
    gene expression and 136
    genes and 136
    interconnected relationship of 134

learning and 136
nature vs. nurture and 137
neural networks and 134
brains 134
    architecture of 141
    doing vs. thinking and 146
    language-learning and 135
    neurogenesis vs. neuroplasticity
        and 168
BRFSS (Behavioral Risk Factor
    Surveillance System) 113

C

caloric deficit 111
caloric needs 105
caloric nutrients 86, 111
calories 86
    burning vs. storing 87
    intake of 52, 55
    restriction of 52, 67, 108
cancer 1, 8, 64–67, 69
    firefighter awareness of 103
    genetics and 114
    genetic vs. lifestyle 20
    lifestyle and 65, 114–115
    nutrients and 90
    nutrition and 114–115
    obesity and 64
    of the digestive system 65
    prevention of 105
    processed food and 98
carbohydrates 86, 91
Cardinal, Bradley 7
cardiovascular disease (CVD) 19, 91, 102
CDC. *See* Centers for Disease Control
    and Prevention (CDC)
cells 65, 70
    cancer 67
    stem 67, 87
    tumor 67
cellular respiration 65
Centers for Disease Control and
    Prevention (CDC) 8, 64, 106,
    113, 116, 131
cerebral blood flow 66
challenges
    fasting and 51
    overcoming 4–7, 191

change. *See* documenting change;
    leadership and change
change assessment model 14–18
    deficiency vs. excess 16–17
    micro vs. macro 17–18
    nature vs. nurture 15–16
CHD (coronary heart disease) 69
chief executive support 220–221
choices
    freedom of 189
    influences on 21
    intentions and 201
    lifestyle 17–18
    restrictions on 13
    right vs. wrong and 201
    spectrums of 17
cholesterol 67
cognitive function 133–134, 168–169
    job performance and 168
community 144, 155
computerized tomography (CT scan) 134
coronary heart disease (CHD) 69
cortisol 69, 142
courage 13
COVID-19 8
creatine phosphate 40
CT scan (computerized tomography) 134
culture
    biological mechanisms of 23
    changing 23
    department 22
CVD. *See* cardiovascular disease (CVD)

D

data and behavior change 210
deaths. *See* fatalities
decay 32
decay signal 49
decision-making 199–200
    control factors and 202
deficiency vs. excess 16–17, 60
    balance and 13
    extremes of 16
depression 66, 138, 143–144, 211
    added sugar and 116
    anxiety and 143
    metabolic effects of 143
    stress and 143
    suicide and 116
diabetes 69, 102
diet. *See also* nutrients
    diversity of 94
    health outcomes and 114
    nutrients and 89
digital minimalism 157
dimensions
    biopsychosocial 4
    human 195–200
dimensions, biopsychosocial 4
disease prevention 7, 64. *See also* cancer;
    heart disease
disincentives 202
DNA 87, 115
    behavior and 211
documenting change 209–210
    data and 209
dopamine 66, 116
Drive 212
Dunbar, Robin 154
Dunbar's number 154
Durant, Will 2

E

eating windows 54
Eat to Beat Cancer 90, 102
Eat to Beat Disease 87
Eat to Live 89, 96
ectomorphs 35–36
education 208
EEG (electroencephalogram) 58
EER (estimated energy
    requirement) 106–108
electroencephalogram (EEG) 58
emergency medical services (EMS) 1,
    188
empathy 158–161
    definition of 158
    individual levels of 159
    limbic system and 144–145
    perspective and 162
    simulation vs. mind theory
        and 159–160
EMS (emergency medical services) 1,
    188

endomorphs 35–36
energy. *See also* estimated energy requirement (EER)
    calculation of required 106–108
    negative balance of 51
    production of 65
    storage of 40
entertainment education 22
Environmental Working Group (EWG) 101
EPA (U.S. Environmental Protection Agency) 101
epigenetics 15, 136
estimated energy requirement (EER) 106–108
eudaemonia 195
eudaemonic motivation 211–214
eudaemonic well-being 132–133
EWG (Environmental Working Group) 101
excess. *See* deficiency vs. excess
exercise 20, 24, 32, 38–41, 42–51, 147–148. *See also* physical activity
    duration of 44–46
    frequency of 43–44
    gear-based activity and 45–47
    glycogen storage and 87
    intensity of 45–48
    means of 51
    medications and 66
    mental disorders and 147–148
    running and 63
    sleep and 68
    tracking frequency of 43
    volume of 42–43
exercise and health outcomes 63–67
    cancer 64–65
    heart attack 63–64
    suicide 65–66
exertion levels 40
experimentation 207

F

failure
    mental rehearsal and 151
    self-talk and 163

fasting 39–41, 51–56. *See also* Human Growth Hormone (HGH)
    alternate day fasting (ADF) 53
    calorie restriction 52
    challenges and 51
    during shift 54
    firefighting and 54–56
    full-day fasting 53
    intermittent 52–54
    mental disorders and 68
    mental health benefits of 68
    mortality rate and 67
    time-restricted feeding (TRF) 52–53
fasting and health outcomes 66
    cancer 67–68
    heart attack 66
    suicide 68–69
fat 40, 67, 86, 91. *See also* triglycerides
    storage of 40, 86
fatalities
    causes of 8–9
    job-related 1
    statistics 8
FDA (Food and Drug Administration) 110
fear network 142
Federal Emergency Management Agency (FEMA) 25
feedback loops 13, 18
feelings 158
FEMA (Federal Emergency Management Administration) 25
Fifth Needs Assessment 24
fight-or-flight response 142
finding balance 60–61
fire academy 188
fire departments
    development cycle of 216
    growth and 219
    introducing change to 218
    major goals of 216
    needs of 15
fire engines 179–180
firefighters
    athleticism of 35–36, 44
    fatalities and 8
    fitness culture of 22
    health and wellness training for 7
    health of 1

obesity and 37
stress and 16
firefighting
cancer and 115
empathy and 159
fire service culture 22
followership 189
food. *See also* nutrition; processed food; raising food
cancer prevention and 115
chemical residue on 101
deficient vs. excessive consumption of 93
groups 92–94
labeling of 99–100
labels 99–100
nonorganic 101
organic 101
pyramid 92
types and colors of 93–95
Food and Drug Administration (FDA) 110
forgiveness 204
fuel 39–40, 48, 86. *See also* fat
types of 40
Fuhrman, Joel 89, 102
full-day fasting 53–54

G

gamma-aminobutyric acid (GABA) 66, 116
gear-based activity (GBA) 45–50, 64
metabolic gears of 47–48
genes
expression of 136
genetically modified organisms (GMOs) 99
genetics 15–16, 136, 199
behavior and 211
cancer and 114
code 15
environmental exposures and 16
inheritance 34–38
mental disorders and 138
physical traits and 16
regulation of 16
GI (Glycemic Index) 102
gluconeogenesis 41

Glycemic Index (GI) 102
glycogen 86
glyphosate 100–101
GMOs (genetically modified organisms) 99
goal-setting 62, 203–204
reassessment and 204
SMART model of 203
group relationships
Dunbar's number and 154
types of 154–155
growth mindset 163
growth signal 49
gut 95

H

habits 105–106, 141, 157
health
behavioral 9–10
definition of 2, 183
importance of 2
mitochondrial 65
optimization of 13, 50
physical 9–10
health and wellness system 219–223
administrative support 223
peer support 223–224
policy development 222
preventative education resources 222
heart attacks 8, 17, 19, 63–64, 66, 68, 170
nutrition and 113–114
heart disease 1, 67, 69. *See also* heart attacks
added-sugar intake and 114
diet and 113
fasting and 66
fatality statistics of 8
processed food and 98
hedonia 195
hedonic motivation 211–214
reward vs. punishment 211
hedonic well-being 132
HGH. *See* Human Growth Hormone (HGH)
Highly Sensitive Persons (HSPs) 140
human dimensions 195–200
Human Genome Project 15

Human Growth Hormone (HGH) 31–34, 69. *See also* exercise; fasting; sleep
  activation of 34
  decline of 32–34, 46
  opportunities for 63
  production of 34, 62–63, 64
  release of 39, 55
human resources and peer support 222–224
hydration 108–110
  calculation of 109–110
  hyponatremia and 109
hypertension 69
hyponatremia 109

I

IAFC (International Association of Fire Chiefs) 46
IAFF/IAFC-WFI. *See* International Association of Fire Fighters and International Association of Fire Chiefs' Wellness Fi
IARC (International Agency for Research on Cancer) 70, 100
ICS (Incident Command System) 217
IF (intermittent fasting) 52–54, 67
imagination and mental rehearsal 150
immune system 87–89
incentives 205
incident command system (ICS) 217
individual knowledge 12
individual responsibility 11–14. *See also* responsibility
  ability to choose and 13–14
  circumstances and 12
  knowing right from wrong and 13
  voluntary choice and 13
inner dialogue 163. *See also* self-talk
intelligence (IQ)
  genetic vs. environmental impact on 139
  learning ability and 139
  life success and 139
intention and mindfulness 148
INTERHEART study 170
intermittent fasting (IF) 52–54, 67

International Agency for Research on Cancer (IARC) 70, 100
International Association of Fire Chiefs (IAFC) 46
International Association of Fire Fighters and International Association of Fire Chiefs' Wellness Fitness Initiative (IAFF/IAFC-WFI) 10, 222, 223
IQ. *See* intelligence (IQ)
isolation 153

J

job-related deaths 1. *See also* line-of-duty deaths (LODDs)
jogging 47. *See also* exercise; physical activity

K

Kandel, Eric 135–137
Keys, Ancel 113
knowledge
  assessing 204
  individual 12, 202
  transfer of 206
knowledge gaps. *See also* leadership: closing the knowledge gap
  identification of 205
  leadership and 189
  methods of reducing 194, 206–207

L

ladder trucks 179–180
LDL (low-density lipoproteins) particles 67
leadership
  definition of 180–181
  diversity in 196
  equation for quantifying 191
  forgiveness and 199
  learning and 198
  love and 199
  management vs. 180–181, 188–189, 192–193, 212
leadership and change 184–196
  achieving goals 184–187
  behavior change 190
  establishing goals 186–187

freedom of choice 189
goal alignment 184
guidance rejection 192
human dimensions of
 interest 194–195
motivation strategies 195
positive vs. negative change 185
reduction of knowledge gaps 186–188, 194
sustaining change over time 190–195
voluntary vs. forced change 187–189
leadership and closing the knowledge
 gap 199–209
 addressing the behavior change
  gap 206–209
 assessing behavior 199–202
 assessing the behavior change
  gap 203–206
 delivering the learning plan 208
 developing a learning plan 206–207
 documenting change 209–210
leadership and human
 dimensions 195–200
 the body dimension 198–199
 diversity of approach 196
 function vs. aesthetic 197
 the mind dimension 198
 positive interactions 196
 resiliency 196
 the soul dimension 199–200
leadership and motivation 210–214
 hedonic vs. eudaemonic
  motivation 211–214
 individual preparedness for
  change 210
 stages of behavior change 210–211
 three main motivators 212–213
leadership and organizational
 change 214–224
 critical components of
  success 220–222
 health and wellness system 219–221
 the individual 215
 organizational systems
  development 217–220
 system design guidance 221
learning 139–140, 204–205
 delivery of 209
 educating vs. 208

experimentation and 207
leadership and 198
observation and 207
unisensory training and 140
learning plans 206–209
 delivery of 208
 development of 207–208
life balance model 165–166
life phases 33
life quality 167
lifestyle 1
 cancer and 114–115
 choices 17–18
 factors 9, 20
 issues 19
limbic resonance 23, 145
 spectrum of 145
limbic system 142, 145
 depression and 143
limits 6
line-of-duty deaths (LODDs) 19, 222
 heart attack and 114
 prevention of 11
Littman-Ovadia, Hadassah 165
liver 40, 41
Li, William 87, 90
LODDs. *See* line-of-duty deaths
 (LODDs)
Lodge, Henry 40
low-density lipoproteins (LDL)
 particles 67

M

macro (the group and you) 17–18, 21–25
 culture 22–23
 cyclical relationship 18
 resources 24–25
 rules 23–24
magnetic resonance imaging (MRI) 134
management 181
 incentives vs. disincentives and 206
management science 189, 212
mastery 61, 213
 self-efficacy theory and 213
MD Anderson Cancer Center, University
 of Texas-Houston 19, 114
meal preparation 90, 103
media usage 66

Mediterranean diet 91
  depression and 116
  heart health and 113
melatonin 69
mental disorders 20, 65, 68, 116
  diagnosis of 132, 143
  genetic component of 138
  obesity and 116–117
  prevention of 66
  suicide and 65–66
mental health 131–134
  cognitive function 133–134
  definition of 131–134
  determination of 138
  improvement of 137
  nature vs. nurture and 138
  optimization of 133–134, 166
  psychological distress 132
  psychological well-being 132–133
mental rehearsal 150–152
  anxiety reduction and 150
mental wellness 131–145
  anxiety and 142
  depression and 143–144
  individuals and groups and 144–145
  intelligence and 139
  quality of engagement and 144–145
  sensory function and 139–140
  thought patterns and 141–142
mesomorphs 35–36
metabolic equivalent of task (MET) 42–47
  scores 44
metabolic gears 47
metabolic syndrome 69
metabolism 67
microbiome 87, 95, 115
  neurotransmitters and 116
micro (you and the cell) 17–20
  cancer 19–20
  cyclical relationship 18
  heart attack 19
  suicide 20
Mifflin-St. Jeor equation 107–108
mind 134, 198
  subjectivity of 134
mindfulness 147–149
  definition of 148–149
  meditation vs. 149
  strategies of 149
  stress reduction and 147
mind theory 160–161
minerals 86
mirror neurons 23. See also limbic system
  relationships and 155
mitochondria 65
  functionality of 67
mitophagy 65
Monsanto 100–101
motivation 210–214, 211–215
  hedonic vs. eudaemonic 211–214
motivators
  autonomy 212
  mastery 213
  purpose 212–213
MRI (magnetic resonance imaging) 134
muscles 40
  glycogen storage in 54
MyPlate model 89, 90, 92–93

N

National Fallen Firefighters Foundation (NFFF) 10, 222, 223
National Fire Protection Association (NFPA) 10, 222
National Institute for Occupational Safety and Health (NIOSH) 64, 115, 222
nature vs. nurture 15
Neff, Kristin 161
negative energy balance 51
nerve growth factor (NGF) 68
neural networks 134
neurogenesis 168
neurological diseases 169
  Alzheimer's 169
  mental wellness and 169
  Parkinson's 169
neurons 141, 146, 168
  mirror 23
neuroplasticity 141, 168
neurotransmitters and mental disorders 116
Newport, Cal 157
NFFF. See National Fallen Firefighters Foundation (NFFF)

NFPA (National Fire Protection Association) 10, 222
NFPA Standards 10
  NFPA 1500 11, 14, 24, 38, 113, 167, 221, 222
  NFPA 1582 46
  NFPA 1583 11, 14, 24, 38, 113, 167, 221, 222
  NFPA 1901 179, 196
NGF (nerve growth factor) 68
night-shift and cancer 70
NIOSH. *See* National Institute for Occupational Safety and Health (NIOSH)
nutrient content 101–104
  Aggregate Nutrient Density Index (ANDI) 102–103
  Eat to Beat Cancer 102
  Glycemic Index (GI) 102
nutrient density 95–105
  databases of 101
  deficiency vs. excess and 103
  food processing and 96–99
  nutrient content and 101–104
  raising food and 99–101
nutrient quantity 105–111
  calculation of 106–108
  habit and 105–106
  hydration calculation and 108–110
  obesity and 106
  supplements and 110–111
nutrients
  balance of 95
  caloric vs. noncaloric 86–87, 89, 91–92, 95
  consumption recommendations of 91
  density of 89–90
  exclusion of 91
  food color and 94
  function of 86–87
  impact of 87–88
  noncaloric 86
  quantity of 90
  raising food and 99
  requirements of 85–86
  variety of 89–90
nutrient variety 90–95
  caloric vs. noncaloric nutrients 91–92
  food groups 92–93
  food types and colors 93–94

nutrition 198
  cancer and 114–115
  heart attack and 113–114
  labels 98
  suicide and 116–117
nutritional psychology 116
nutritional wellness 85–90
  nutrient density 89–90
  nutrient function 86–87
  nutrient impact 87–88
  nutrient quantity 90
  nutrient requirements 85–86
  nutrient variety 89–90

O

obesity 37, 66, 69
  cancer and 64, 106, 115
  FTO gene and 36–37
  heart health risks and 114
  mental disorders and 116–117
  rates of 106, 114, 116
observation 207
Occupational Safety and Health Administration (OSHA) 10, 64, 221
Ohsumi, Yoshinori 67
opportunities 164
organizational leadership 215
  addressing the change gap and 218
  development cycle of 217
  resiliency and 219
organizational performance 215–216
organizational responsibility 11–14
organizational systems development 217–220
  chief executive support for 220–221
  critical components of 220–222
  health and wellness example 219–222
  human resources and 222–224
  incident command system (ICS) and 217
  invention vs. adoption of systems 218
  maintenance 217
  system design 221–222
organizations 214
  composition of 215
  management and 214
  three dimensions of 214–215

OSHA. *See* Occupational Safety and Health Administration (OSHA)
overcoming challenges 4–6, 191
oxytocin 153

P

Parkinson's disease 169
patient interaction 208
peer support 223–224
performance improvement 215
personal protective equipment (PPE) 11
perspective 162
physical activity 38, 45, 106, 198. *See also* exercise
   jogging 47
   spectrum of 38, 60
physical fitness, assessing 61–63
physical wellness 31–41
   exercise and 38
   fasting and 39–40
   genetic inheritance and 34–38
   human growth hormone decline and 32–34
   sleep and 41
   stress response system and 31–32
phytochemicals 86
Pink, Daniel 195, 212
placebo effect 170–171
positive thinking 171
post-traumatic stress disorder (PTSD) 148
potential 212
PPE (personal protective equipment) 11
prefrontal cortex 143, 168
processed food 96–99
   additives and 98
   cancer and 98
   comorbidities and 114
   depression and 116
   heart disease and 98
   natural alternatives vs. 97–98
   types of 96–97
protein 86, 91
PSOB (Public Safety Officer's Benefit Act) 10
psychological distress 132
   mental disorders and 132
   stressors and 132

psychological well-being 132–133
   hedonic vs. eudaemonic 132–133
   illness and 133
PTSD (post-traumatic stress disorder) 148
Public Safety Officers' Benefit Act (PSOB) 10
purpose 213–214
   utility theory of value and 213

Q

quality of life 167

R

raising food 99–101, 103
   carcinogens and 101
   nutrient content and 99
rapid eye movement (REM) sleep 56–57
Ratey, John 66, 68
rationality 183, 200
relationship optimization 155–159, 162–167
   developing routines and 157
   intentionality with technology and 156–157
   perspective change and 162
   strengthening connections and 155–156
   supportive self-talk and 163–164
   SWOT analysis and 163–165
relationships 144, 153–155
   commonalities and 155
   empathy and 144
   health benefits of 153
   individual vs. group 154–155
   physical distance and 155
   routine and 158
   shared experiences and 155
   SWOT analysis and 165
REM (rapid eye movement sleep) 56–57
reserves 4
resilience 68, 147
   baselines of 5
   bouncing back and 3–4
   definition of 3–5
   equation for calculating 5
   growing 6–7
   leadership and 193, 196

organizational leadership and 219
overcoming challenges and 4–5
reserves and 4
resources 24–25
responsibility 202. *See also* individual responsibility
    assessment 12–14
    organizational 11–14
resting metabolic rate (RMR) 106–107
    Mifflin-St. Jeor equation for calculating 107–108
Roundup 100–102
routines 157
rules 23–24
running 64. *See also* exercise; physical activity
    cancer mortality and 65
    cardiac mortality and 64

S

SCBA (self-contained breathing apparatus) 46
schizophrenia 138, 211
s curve 60
sedentary state 38, 44, 66
selective serotonin reuptake inhibitors (SSRIs) 68
self-compassion 158, 161–162
    acknowledgement of pain and 161
    self-pity vs. 161
self-confidence 151
self-contained breathing apparatus (SCBA) 46
self-criticism 162
self-efficacy theory 213
self-support 163
self-talk 163–164
    neuroplasticity and 169
senses 139
    genetics and 139
    sensory continuum 140
serotonin 66, 116
Seven Countries Study 113
Shapiro, Shauna 148
simulation theory 159
Sixteen Firefighter Life Safety Initiatives 151, 222

sleep 41, 56–60, 66
    cycles of 58–60
    disruption of 58, 69–70
    fasting and 68
    improvement methods for 57–58
    lack of 56
    physical activity and 68
    spectrum of 58–59
    stages of 42, 56–58
    suggested amount of 60
    volume of 57
sleep and health outcomes 68–70
    cancer 69
    heart attack 68
    suicide 70–71
SMART model 203
social connection
    individuals vs. groups and 144–145, 152–159
    quality of engagement and 144–145, 158–166
social interaction and health 154. *See also* relationships
social shelter 157
social support 24
solitude 144, 152–154
    benefits of 152
    finding a balance of 153
    isolation vs. 153
    psychological well-being and 144
    SWOT analysis and 165
soul 183, 199
Spark 66
spirit 183
spirituality 183
SSRIs (selective serotonin reuptake inhibitors) 68
states
    active 38
    sedentary 38, 44, 66, 111
stem cells 115
    system 67, 87
strengths 164
stress 16, 170. *See also* cortisol; stress response
    anxiety and 150
    cancer and 170
    heart disease and 170
    hormone 142

stress (*continued*)
   intensity of 31–32
   sleep and 70
stressors 132, 150
stress response 32, 142
   mindfulness and 149
   system 31–32
sugar 40, 86, 114
   storage of 40
   types of 98
sugar, added 116
suicide 1, 20, 65–70
   fatality statistics of 8
   mental disorders and 65–66
   nutrition and 116–117
   relationships and 170
   risk factors of 116
   risk lifestyles and 66
   sleep and 71
   thought patterns and 170
supplements 110–111
   hormone production and 110
   types of 110
support teams 223
SWOT analysis 163–165
system design 220–221
   national guidance and 221–222
   process support and 220
systems development. *See* organizational systems development

T

teaching vs. leading 208
technology
   apps and 157
   relationships and 156
   solitude and 156
thermic effect of feeding (TEF) 106
thought patterns 141–142, 146–152
   doing vs. being and 146–149
   exercise and 147–148
   habits and 141
   mental rehearsal and 150–152
   mindfulness and 147–149
   neural pathways and 141
   optimization of 168
threats 165
time-restricted feeding (TRF) 52–53

Transtheoretical Model of Behavior Change (TTM) 210–211
TRF (time-restricted feeding) 52
triglycerides 40, 48, 67, 86. *See also* fat
TTM (Transtheoretical Model of Behavior Change) 210–211

U

Universal Health Atlas 102
University of Sydney Glycemic Index 102. *See also* glycemic index
University of Texas-Houston 19
U.S. Department of Agriculture (USDA) 89
U.S. Environmental Protection Agency (EPA) 101
utility theory of value 213

V

value 213
virtue 13, 211
virtuous middle 43, 51, 201
vitamins 86–88. *See also* nutritional wellness

W

water 86, 88, 108–109
weaknesses 164
weigh-in rules 24
weight 90
   assessing 204
   limits 37
   maintenance 90
weight gain 90
   routine and 105
weight loss 90, 108, 111. *See* fasting
   high-impact behaviors and 111
   weigh-in rules and 24
weight management 111–113
   accountability chart 111–112
wellness. *See also* wellness education
   balance of 17
   champions 220, 223
   deficit of 17
   definition of 2
   efforts of 3

Index   255

habits of 7, 17
importance of 2
types of 16
wellness education 7–11, 24
    big three and 8
    lifestyle and 9–10
    preventative resources and 222

standards of 10
World Health Organization (WHO) 116, 167
writing 165

Y

Younger Next Year 40, 49

ABOUT THE AUTHORS

Gamaliel Baer is a firefighter/EMT and special operator for Howard County (MD) Fire and Rescue. He currently serves as a U.S. Coast Guard Reserve officer and is a volunteer for the National Fallen Firefighters Foundation. He is credentialed by the Center for Public Safety Excellence (CPSE) as a Chief Training Officer (CTO), and by the American College of Exercise (ACE) as a certified health coach and a certified personal trainer. Dr. Baer serves as adjunct faculty at Johns Hopkins University where he teaches Leading and Managing Change for the Master of Science in Organizational Leadership.

Dr. Baer earned his Bachelor of Science in Marketing and International Business at the University of Maryland, College Park. He earned his Master of Science in Management from Johns Hopkins University, and he earned a Doctor of Education in Organizational Leadership from the University of Southern California. He lives in Howard County, Maryland, with his wife and four children.

David Schary is an associate professor of Exercise Science at Winthrop University where he focuses on improving athlete mental health and performance. In addition, Dr. Schary consults with Winthrop Athletics, the U.S. Forest Service, and the National Fallen Firefighters Foundation. Prior to academia, Dr. Schary coached rowing at the high school and collegiate levels.

Dr. Schary completed his PhD in Exercise and Sport Science with a concentration in Exercise and Sport Psychology and Master of Public Health with a concentration in Biostatistics at Oregon State University. He also holds a Master of Science in Exercise and Sport Studies from Smith College and a Bachelor of Arts in Sociology from the University of California, Davis. Currently, Dr. Schary is enrolled in Winthrop University's Clinical Mental Health Counseling program.